AF675667

Design Principles of Autonomous Systems

This book explains unmanned systems, including Unmanned Aerial Vehicles (UAVs), Autonomous Underwater Vehicles (AUVs), and Unmanned Ground Vehicles (UGVs). It also details system and subsystem compositions and their basic design. The components/subcomponents, employed payloads and sensors, and communication systems used to compose the autonomous system are discussed. It examines generic applications of these unmanned systems including specific missions for which they are employed. Other topics like swarm of drone, anti-drone system, and some algorithms used in navigation and communication of the drone are also discussed.

Some key features:

- Helps readers understand nuances of autonomous systems on land, in air, and in sea.
- Explains pertinent design principles, sensors, and communication system.
- Lays the foundation for crafting, designing, and deploying autonomous systems for different applications.
- Reviews algorithms, computing, control and technology stacks required for design of UAS.
- Discusses software and navigation aspects of autonomous systems with the concepts of SLAM.

This book is aimed at graduate students and researchers in mechatronics engineering, systems and sensors, systems and control, and aerospace engineering.

Design Principles of Autonomous Systems

UAV, UGV, and AUV

Narayan Panigrahi and
Smita Tripathy

CRC Press
Taylor & Francis Group
Boca Raton London New York

CRC Press is an imprint of the
Taylor & Francis Group, an **informa** business

First edition published 2026
by CRC Press
2385 NW Executive Center Drive, Suite 320, Boca Raton FL 33431

and by CRC Press
4 Park Square, Milton Park, Abingdon, Oxon, OX14 4RN

CRC Press is an imprint of Taylor & Francis Group, LLC

© 2026 Narayan Panigrahi and Smita Tripathy

Reasonable efforts have been made to publish reliable data and information, but the author and publisher cannot assume responsibility for the validity of all materials or the consequences of their use. The authors and publishers have attempted to trace the copyright holders of all material reproduced in this publication and apologize to copyright holders if permission to publish in this form has not been obtained. If any copyright material has not been acknowledged please write and let us know so we may rectify in any future reprint.

Except as permitted under U.S. Copyright Law, no part of this book may be reprinted, reproduced, transmitted, or utilized in any form by any electronic, mechanical, or other means, now known or hereafter invented, including photocopying, microfilming, and recording, or in any information storage or retrieval system, without written permission from the publishers.

For permission to photocopy or use material electronically from this work, access www.copyright.com or contact the Copyright Clearance Center, Inc. (CCC), 222 Rosewood Drive, Danvers, MA 01923, 978-750-8400. For works that are not available on CCC please contact mpkbookspermissions@tandf.co.uk

Trademark notice: Product or corporate names may be trademarks or registered trademarks and are used only for identification and explanation without intent to infringe.

ISBN: 9781032452678 (hbk)
ISBN: 9781032558769 (pbk)
ISBN: 9781003432685 (ebk)

DOI: 10.1201/9781003432685

Typeset in Times
by Newgen Publishing UK

Contents

Preface vii
About the authors ix

1 Introduction to Unmanned Autonomous Systems (UAS) 1

2 Crafting an Unmanned Aerial Vehicle (UAV) 21

3 Designing the fuselage, wing, propulsion, and empennage of a fixed-wing UAV 67

4 Application-specific UAVs and their requirements 81

5 Classifications of different UAVs 99

6 Counter-UAS systems 122

7 Design criteria of a UAV for photogrammetry and ISTAR applications 156

8 Autonomous Underwater Vehicles (AUVs) 168

9 Essential design concepts of AUVs 196

10 Unmanned Ground Vehicles (UGVs) 209

11 Crafting a UGV 225

12 Algorithms, control, and software used for UAS missions 254

Index 277

Preface

There is a great euphoria amongst academia and industry regarding research and development in the field of Unmanned Autonomous System (UAS). This has resulted in a surge of research literature in the public domain pertaining to development of UAS and its allied technologies. There is an increasing demand for materials for teaching and learning about this multidisciplinary applied science. This book is authored with an intention to fill the gap in the pedagogy of UAS in general and autonomous systems in particular.

UAS is a collaborative system encompassing many technologies, such as avionics, propulsion system, sensing, navigation, cognitive science, control system, and digital communication, to name a few. A UAS manifests itself in different forms, such as Unmanned Ground Vehicles (UGVs), Autonomous Underwater Vehicles (AUVs), and Unmanned Aerial Vehicles (UAVs) operating on the ground, under the sea, and in the air. They are deployed for varying missions by different application communities. In this book we have not taken into consideration autonomous sea-surface vehicle and autonomous vehicles for the exploration of space and planet surfaces such as Mars and the Moon rover.

In this book we have discussed the design aspects of UAVs, UGVs, and AUVs and their various from factors, their classification based on different design criteria and operating features. The aspect of autonomy is discussed in the form of sensing computing and control loop which is discussed in Chapter 12. How the control loop manifests in the form of software and hardware combination is also discussed. Size, Weight, Area and Power (SWAP) is an import design criterion of the UAS. Therefore, keeping in view its operation and application, various SWAP constraints are applied to the designing of different UAS systems. We have discussed and classified the UAS systems based on applications, SWAP constraints, and its operation parameters in a separate chapter.

The first chapter of this book discusses the general concepts of UASs and the taxonomy of UASs. Chapters 2–5 discuss UAV design and applications. Chapter 6 considers various counter-UAS technologies and systems. Chapters 7–9 investigate the design and applications of AUVs. Chapters 10 and 11 are devoted to the design and applications of the UGV. Finally, Chapter 12 discusses the software and algorithms used in the UAS for achieving autonomy. Each chapter has a Question and Answer section to test the effectiveness of the pedagogy among readers of this book. We sincerely hope this will serve its intended purpose of teaching and learning the concepts of UASs and allied subjects.

About the authors

Dr Narayan Panigrahi heads a team of scientists at the Center for Artificial Intelligence and Robotics, a premier laboratory of the Defence Research and Development Organization, India. He is engaged in building capacity and capability in the research and development of Geographic Information System (GIS) and its applications in different domains and is also pursuing his research for the design and development of indigenous geospatial and spatio-temporal algorithms. He has been instrumental in designing and developing Indigenous GIS for military applications and GIS for cross-domain applications in terrain-bathymetry and space. He is the recipient of many awards, including the Best Graduate & Governor's Gold Medal of Berhampur University (1987), Best Research Award by INRIA France (2012), and the Agni Award for Self-Reliance (2019). He is a recipient of the National Geospatial Award for excellence (2019) and National Geomatics Award for excellence (2023).

Smita Tripathy joined as a scientist in the Light Combat Aircraft (LCA) programme of the Aeronautical Development Agency, Ministry of Defence, India, in 1998. Since then, she has been engaged in the research and development of avionic systems of LCA. She has contributed to the design and development of avionics system, and her primary interest lies in the design and development of avionic systems and subsystems of LCA.

Chapter 1

Introduction to Unmanned Autonomous Systems (UAS)

OVERVIEW

This chapter introduces the concepts and taxonomy of the class of vehicles known as Unmanned Autonomous Systems (UAS) (Valavanis and Vachtsevanos, 2015). This class of vehicles are also known as pilotless vehicles, or autonomous vehicles, of which the Unmanned Aerial Vehicle (UAV) is quite popular among a large community of users, media, and news. This chapter first gives a basic definition and taxonomy of the broad class of systems known as UAS. It discusses the most common prototype unmanned system known as UAV. To kindle interest among the first-time readers of this book, some popular applications of UAVs are discussed. In addition, the chapter gives a pictorial description of the organization of the book and its chapters in the form of a bubble diagram.

The chapter begins by answering fundamental questions such as "What is a UAS?" and "What is a UAV?" while also providing a historical overview of unmanned systems' development. The evolution of unmanned vehicles is traced through bibliographic descriptions. The text also addresses common misconceptions and confusion surrounding unmanned systems, such as whether a missile or kinetic weapon qualifies as a UAV and whether a drone and a UAV are the same thing. To resolve these ambiguities, the chapter provides clear definitions of the three terms: drone, UAV, and UAS.

1.1 WHAT IS A UAS?

An Unmanned Autonomous System (UAS) is a sophisticated system-of-systems that comprises various well-coordinated subsystems which work together to achieve mission objectives. The subsystems that make up a UAS include the ground control system, telemetry subsystem, mobile data storage and analysis centre, sensor payload, navigation system, and the Unmanned Aerial Vehicle (UAV) itself.

The ground control system is responsible for monitoring and controlling the UAV's movements during the mission. The telemetry subsystem enables

DOI: 10.1201/9781003432685-1

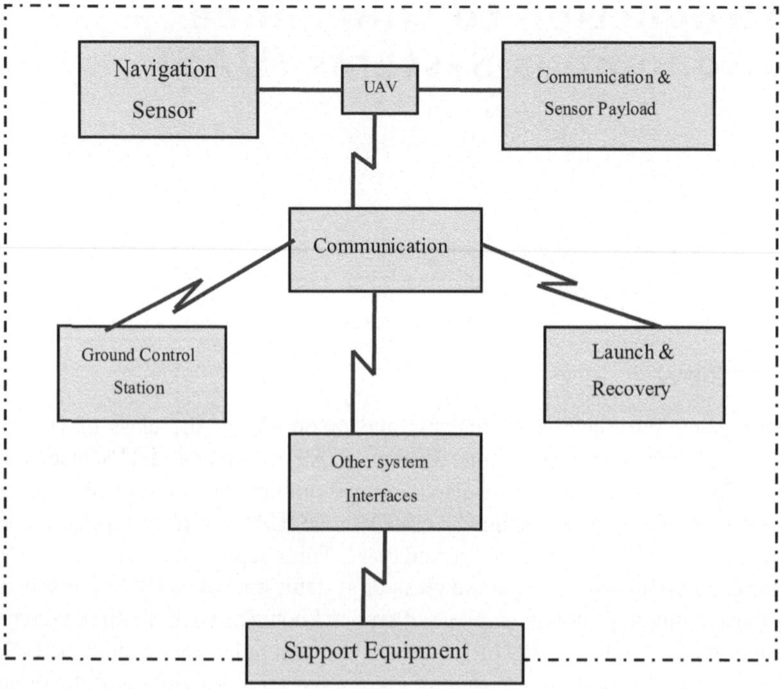

Figure 1.1 Abstract block diagram of a UAS.

communication between the ground control and the UAV. The mobile data storage and analysis centre collects, stores, and analyses data captured during the mission. The sensor payload and navigation system mounted on the UAV enable the UAS to collect relevant information and navigate through the mission environment.

It is important to note that the UAV is a critical component of the overall UAS and cannot function without the other subsystems. These subsystems include the ground control system, transmission systems, Global Positioning System (GPS) module, and software. In essence, the UAS is the complete package that makes a UAV work.

To simplify, we can say that a UAS is a complex system that includes multiple subsystems working together to accomplish a mission. The UAV is a critical part of this system, and without the other subsystems, it would not be able to function Given below is a generic block diagram (Figure 1.1) of a UAS with its systems and subsystems interconnected through communication carrying different command, control, and data between various modules. A pictorial realization of the UAS with its various elements and their interrelation is depicted in the Figure 1.2.

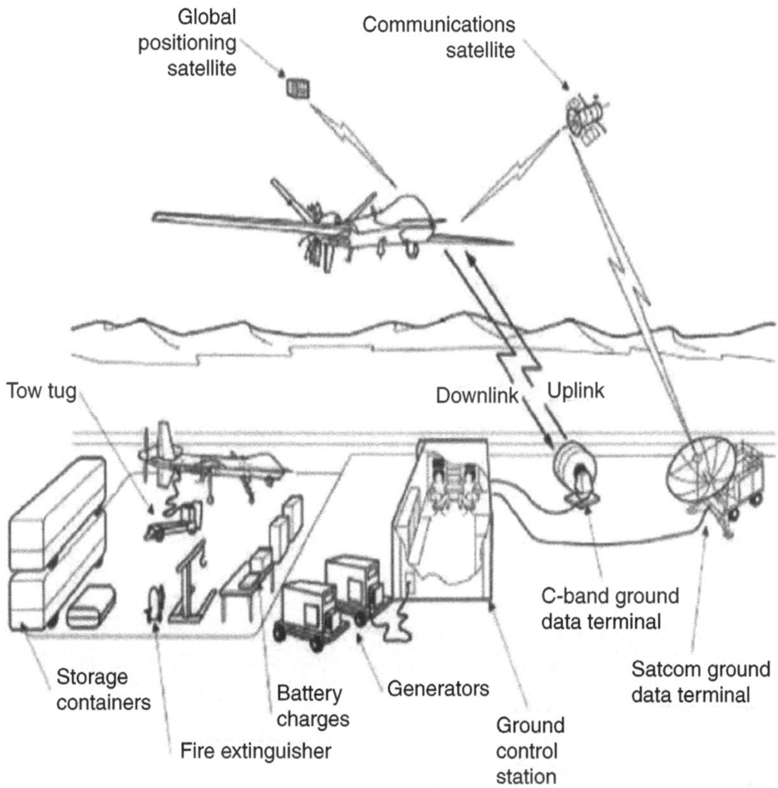

Figure 1.2 Pictorial diagram of UAS with associated systems and subsystems.

1.2 WHAT IS A UAV?

A UAV is a specific derived class of aerial vehicle from the generic class of systems known as UAS. UAVs give an understanding that these are a class of aircraft, which can fly without the presence of an on-board pilot. Some of the common terms used for UAV are drone, UAS, and remotely piloted aircraft (RPA). These terms mean different for a professional in the field and designate different unmanned craft (Vergouw et al., 2016).

UAVs are quite often expected to have autonomous flight capabilities, which is not the case with drones. Drones can be controlled by on-ground pilots. In short, all UAVs can be called drones, but not all drones can be called UAVs. UAS is a term collectively used to define the on-ground control element along with the

UAV, which is an integral part of it. An RPA, which received its acronym from the International Civil Aviation Organization (ICAO), is referred to as a UAV with complex control elements and groundbreaking innovations. Despite this, however, UAVs and RPAs can be interchanged.

1.3 WHAT IS A DRONE?

"Drone" is the most commonly used term to refer to any unmanned aerial vehicle, mostly because it is a term that the media knows will catch the attention of casual viewers. From a technical standpoint, a drone can refer to any vehicle – even those that travel in water or on land – that can travel autonomously.

In many ways, drone is a term that can be used as a non-technical catch-all for both UAVs and UASs. The term can be used to refer to both fixed-wing and multi-copter models, although the latter type has become much more common. In modern times, the definition of drone has expanded to include vehicles that can be controlled remotely, in addition to those that are capable of autonomous movement. Thus, a simpler way of defining a drone is simply as any vehicle that can operate without a driver or pilot inside it.

Nowadays, drone is a good word to use for popular media such as movies and television, but may be insufficiently specific for technical conversations. In short, all UAVs can be called drones but not all drones cannot be called UAVs. Similarly, a UAV is part of a UAS and is a subsystem of the complete system (Yaacoub et al., 2020).

1.4 DIFFERENCE BETWEEN A UAS, A UAV, AND A DRONE?

Drones have been in the news a lot in the past couple of years, both for good and bad reasons. Some drones have been used to fight forest fires, aid in search and rescue operations, or help law enforcement at practically zero risk. On the other hand, there are also reports of drones becoming a nuisance at airport runways, being used as drug couriers, or spying on unsuspecting civilians (Mohsan et al., 2022). Figure 1.3 (a), (b) ,and (c) depicts a skeletal picture of a drone with its various components.

All the talk about drones raises the questions: What exactly is a drone? How is a drone different from the less common but still often mentioned UAV and UAS?

1.5 WHAT'S NEXT FOR THE UAV LEXICON?

The field of drones, or UAVs, is continuously evolving, and we can expect the UAV lexicon to undergo further development in the coming years. Currently, a

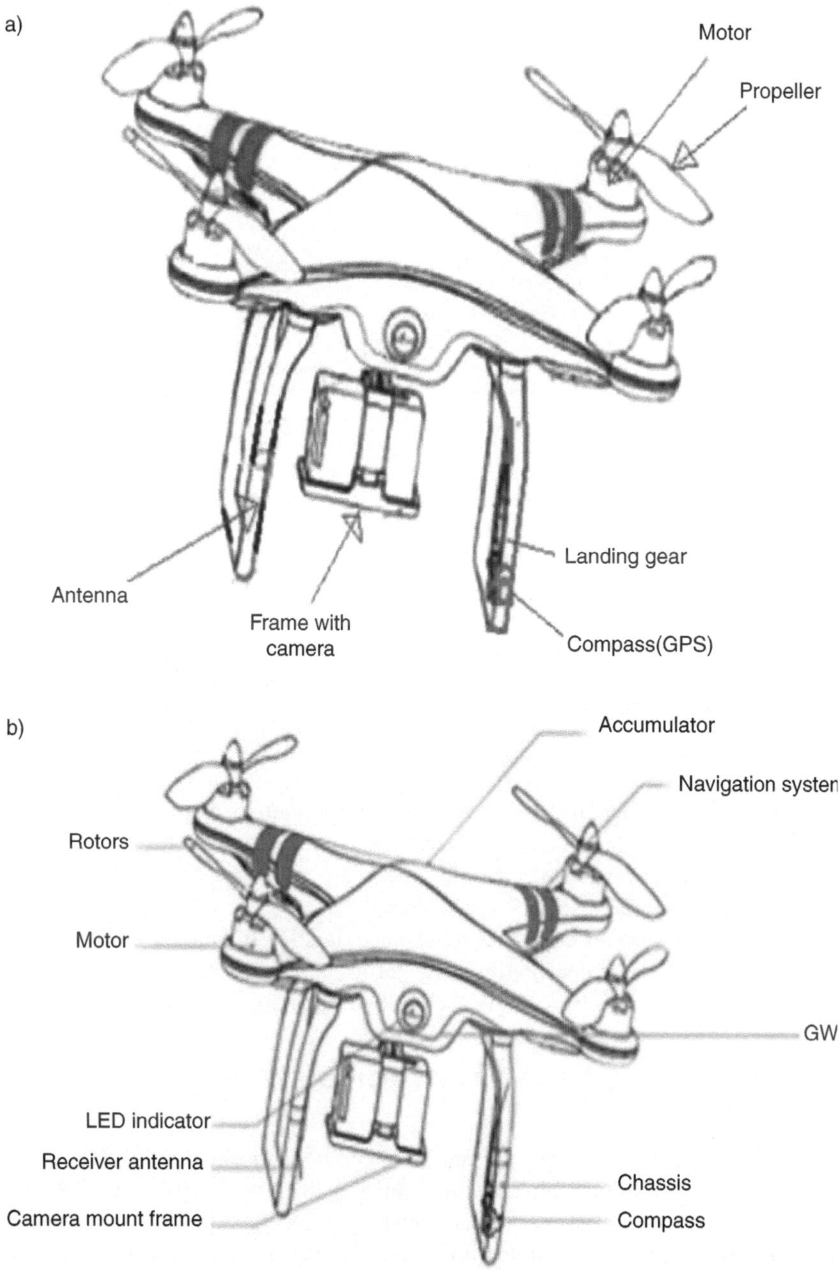

Figure 1.3(a)(b) Skeletal diagram of UAV with its different parts.

c)

Figure 1.3(c) A real picture of a drone.

distinction is being made between UAVs and RPAs in some references. This differentiation aims to highlight high-level drones that require extensive training or expertise compared to recreational models used for casual flying.

However, the widespread adoption of the term "RPA" has not yet occurred, and UAV and RPA are still commonly used interchangeably. In everyday conversations, articles, and news reports, the term "drone" suffices to convey the intended meaning. For most people, the term "drone" brings to mind the image of a standard multi-copter drone, which aligns with their prevalent popularity.

Nevertheless, in technical contexts, it is important to distinguish between UAV and UAS. While UAV refers specifically to the drone itself, UAS encompasses the entire ecosystem, including hardware, software, skills, and human input, that enables the operation of the drone.

As the field of drones continues to advance, it is likely that the UAV lexicon will further evolve, incorporating new terms and distinctions to accurately represent the diverse range of UAV technologies and applications.

1.6 EMERGENCE OF THE UAV, FOR DIFFERENT APPLICATIONS

Early examples of UAV concepts can be traced back to as early as 1849 when Austria attacked Venice using unmanned balloons filled with explosives. Another example was a configuration of quadcopters that appeared in 1907, although a pilot was still required to control the flight. However, it was during the First

World War that the first pilotless vehicles were developed in both Britain and the USA, primarily for use as aerial targets (Wu, 2017).

During the Second World War, unmanned aircraft were equipped with radio controllers and launch mechanisms, although their range capabilities were limited. These early UAVs utilized a guidance system that employed a barometer to regulate the aircraft's speed and altitude, along with a vane anemometer to measure the distance travelled. Additionally, nose-mounted cameras were controlled via radio signals transmitted to a television screen in a manned aircraft, providing initial guidance and control of the system.

The 1960s marked a significant development in UAV technology with the introduction of transistor technology, making miniature remote-controlled (RC) controllers accessible to civilian users. This led to the emergence of UAVs for recreational purposes, with RC plane kits becoming available to hobbyists. Meanwhile, the military continued to utilize UAVs with still cameras for reconnaissance missions in enemy territories.

In the 1970s, UAVs were further advanced for long-range, high-altitude reconnaissance, providing real-time information about the battlefield. This era saw the evolution of UAVs into sophisticated systems capable of operating at extended ranges and high altitudes. By the 1990s, with the advent of GPS and satellite communication, UAVs were deployed as high-altitude long-endurance (HALE) and medium-altitude long-endurance (MALE) systems, enabling them to remain airborne for extended periods and maintain secure communication links while avoiding ground detection systems.

From the early 2000s to the 2010s, drones experienced significant advancements in commercial applications. The usage of drones for delivery purposes gained prominence, resulting in a surge in demand for drone permits. The integration of RC controllers with smartphones opened up new possibilities for utilizing drones in photography and videography, leading to an increase in their popularity among consumers. Moreover, the decreasing prices of components such as microcontrollers, accelerometers, and camera sensors facilitated the development of DIY (Do-it-Yourself) drones used by various sectors, including the police and fire services.

Overall, the history of UAVs has witnessed remarkable progress in terms of their capabilities, applications, and accessibility. From their early military deployments to their widespread adoption in civilian sectors, drones have undergone significant technological advancements, shaping their current role as versatile tools for a range of purposes.

Present-day drone innovation has found applications in sectors such as agriculture, construction and mining, insurance, and telecommunications. With

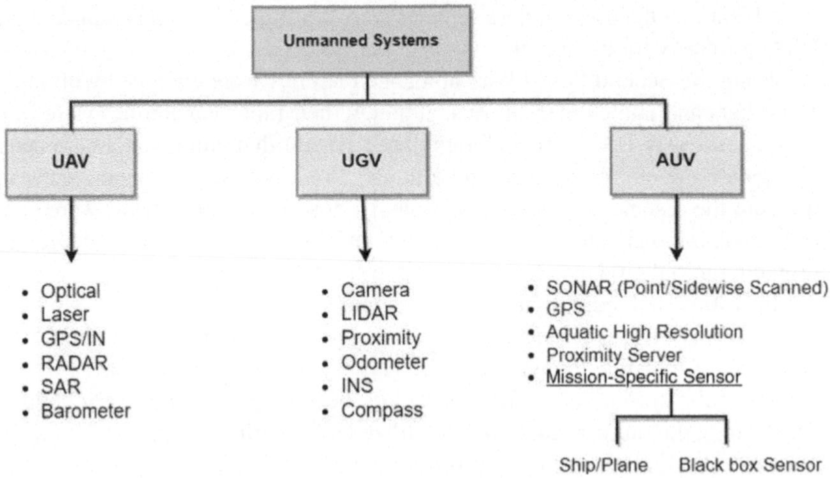

Figure 1.4 Different sensors employed in different category of UASs.

the ability to miniaturize components and improved communication systems, drones are also being used as emergency medical assistants, and micro-drones are being used by the military for surveillance purposes.

With the establishment of 5G networks in the recent past, theoretical designs of a swarm of drones being used to create heterogeneous networks (HetNets) are being proposed. There have also been proposed improvements in battery technologies and ideas to make use of drones as energy harvesting devices (Natalie, 2005; Hildmann, 2019) (Figure 1.4).

1.7 INTRODUCTION TO UNMANNED AUTONOMOUS SYSTEMS (UAS)

The selection of a UAS configuration depends on the specific speed requirements of the mission. Different configurations, such as lighter-than-air aircraft, fixed-wing aircraft, rotary-wing aircraft, or convertible aircraft, are used based on these requirements.

For long-endurance and long-range military surveillance missions, a high-aspect ratio fixed-wing aircraft operating at high altitude is typically used. These aircraft require long paved runways to achieve the necessary lift-off speed due to the high wing-loading required for low aerodynamic drag.

In the case of Unmanned Combat Air Vehicles (UCAVs) that operate at high speeds, they often have low-aspect ratio wings and can take off from long runways or be air-launched.

Many civilian applications of UAVs require the aircraft to fly at speeds lower than 50 kt (70 km/hr) for a significant portion of the mission. Some applications may require hovering capability, such as power line inspection, while others greatly benefit from it, such as incident control by police and fire services.

Certain military roles, such as naval decoying, army NBC monitoring, laser target designation, air force base security, and the detection and elimination of unexploded bombs (UXB), require or benefit from the ability to hover or fly at very slow speeds.

In scenarios where operations from off-board ships or restricted sites are necessary, vertical take-off and landing capability becomes advantageous. For operations with a top speed requirement of no more than 150 kt (210 km/hr), a helicopter configuration offers efficient hover and low-speed performance. Helicopters are also less sensitive to air turbulence compared to other aircraft types due to their high rotor-blade loading.

Compounded helicopter configurations, which add wings and/or a propulsive system, allow helicopters to achieve higher speeds. The addition of wings can enable a helicopter to reach speeds over 200 kt, while fully compounded helicopters have achieved speeds exceeding 300 kt (550 km/hr). However, these configurations often come at the cost of reduced payload and endurance.

Convertible aircraft configurations aim to strike a viable compromise between vertical take-off and landing capability and long endurance. They lift off with horizontal rotors and then tilt them into a vertical plane to become propellers for cruise flight, with the aircraft's weight borne by wings. However, convertible configurations typically suffer from a payload weight penalty compared to helicopters or fixed-wing aircraft.

Another interesting rotary-wing configuration is the autogyro, which aims to reduce complexity by eliminating the transmission system found in helicopters. Although autogyros cannot hover, they can fly at considerably slower speeds than fixed-wing aircraft. For more detailed information on these different aircraft configurations, refer to Chapter 3.

By considering the specific speed requirements and operational needs, different UAS configurations can be chosen to effectively fulfil a wide range of mission objectives.

1.7.1 The navigation systems

In the context of UAS, ensuring real-time and accurate positional information is crucial for safe and efficient operation. Traditionally, this was achieved through complex and expensive Inertial Navigation Systems (INS) that required frequent updates from ground stations via communication links.

However, the availability of GPS has revolutionized UAS navigation. GPS is a space-based navigation system that provides continuous and accurate positioning information to users worldwide. By utilizing a network of Earth-orbiting satellites, GPS receivers on board UASs can determine their location and velocity.

Compared to INS, GPS offers several advantages. GPS receivers are lighter, more compact, and more affordable, making them suitable for integration into UAS platforms. They provide continuous updates, enabling UAS operators to have real-time information about the aircraft's location. Furthermore, the accuracy of GPS can be further enhanced through Differential GPS (DGPS).

DGPS utilizes a network of ground-based reference stations that transmit correction signals to GPS receivers, improving their positional accuracy. This is especially beneficial in areas with weak GPS signals or obstructions like tall buildings or mountains, as it helps overcome signal degradation.

While GPS and DGPS are highly effective, it is essential to have backup options in case GPS signals are unavailable or blocked. In such situations, alternative navigation methods like visual navigation or inertial navigation can be employed. Visual navigation relies on on-board cameras or sensors to detect and track landmarks or features, while inertial navigation systems use accelerometers and gyroscopes to estimate position based on the aircraft's motion.

These fallback options can be particularly useful when communication between the UAS and ground station is continuous, or when autonomous flight is not required. By considering multiple navigation approaches and having redundancy in place, UAS operators can ensure reliable and resilient navigation capabilities for a variety of operational scenarios.

1.7.1.1 Radar tracking

Radar tracking involves fitting the UAS with a transponder that responds to radar signals emitted from the Control Station (CS). This allows the CS to visualize the UAS position on a radar display, providing information about its bearing and range. Radar tracking is a reliable method for tracking UAS, as it leverages the use of radar technology to precisely determine the UAS's location. It is particularly useful in scenarios where visual tracking is challenging, such as during adverse weather conditions or in areas with limited visibility.

1.7.1.2 Radio tracking

Radio tracking relies on tracking the radio signal transmitted from the UAS to the CS. The CS determines the bearing of the UAS by tracking the direction of the radio signal. Additionally, the range of the UAS is determined by calculating the time taken for a coded signal to travel between the aircraft and the CS.

Radio tracking provides valuable information about the UAS's position and is commonly used in UAS operations. It offers a reliable and efficient means of tracking the UAS, particularly in situations where visual tracking (Zhang, 2019) may not be feasible.

1.7.1.3 Direct reckoning

Direct reckoning involves calculating the UAS's position by integrating velocity vectors and elapsed time. This method relies on precise measurement of the UAS's velocity and the time elapsed since its departure. In cases where the UAS is equipped with a television camera for ground surveying, its position can be further confirmed by comparing visible geographical features with their known positions on a map. The integration of velocity vectors and time allows for the determination of the UAS's position in real time. Direct reckoning can be highly effective when combined with other navigation techniques, providing accurate and continuous positioning information (Kunhoth et al., 2020).

In the interest of ease of operation and efficiency, it is desirable for UAS navigation systems to be as automatic and autonomous as possible. These tracking methods, whether through radar, radio, or direct reckoning, offer reliable ways to monitor and track the UAS during its mission, ensuring effective control and safe operation.

1.7.2 Launch, recovery, and retrieval equipment

1.7.2.1 Launch equipment

Launch equipment is required for those air vehicles which do not have a vertical flight capability, nor have access to a runway of suitable surface and length. This usually takes the form of a ramp along which the aircraft is accelerated on a trolley, propelled by a system of rubber bungees, by compressed air or by rocket, until the aircraft has reached an airspeed at which it can sustain airborne flight.

1.7.2.2 Why unmanned aircraft?

Unmanned aircraft will only exist if they offer advantages compared with manned aircraft (Kardasz, 2016). An aircraft system is designed from the outset to perform a particular role or roles. The designer must decide the type of aircraft most suited to perform the role(s) and, in particular, whether the role(s) may be better achieved with a manned or unmanned solution. In other words, it is impossible to conclude that UAVs always have an advantage or disadvantage compared with manned aircraft systems. It depends vitally on what the task is. An old military adage (which also applies to civilian use) links the use of UAVs to roles which are Dull, Dirty, or Dangerous (D3). There is much truth in that, but it does not go far enough. To D3 should be added covert, diplomatic, research, and environmentally critical roles. In addition, the economics of operation are often to the advantage of the UAV.

1.7.3 Dull roles

Military and civilian applications such as extended surveillance can be a dulling experience for aircrew, with many hours spent on watch without relief, and can lead to a loss of concentration and therefore loss of mission effectiveness. The UAV, with high-resolution colour video, low-light-level television, thermal UAS imaging cameras, or radar scanning, can be more effective as well as cheaper to operate in such roles. The ground-based operators can be readily relieved in a shift-work pattern.

1.7.4 Dirty roles

Applicable to both civilian and military applications, monitoring the environment for nuclear, chemical, and biological contamination puts human operators at unnecessary at risk. Therefore, unmanned crafts are employed to measure and map the contaminated environment so that subsequent detoxification of the aircraft is easier in the case of the UAV. Crop-spraying with toxic chemicals is another dirty role which now is conducted very successfully by UAVs (Restas et al., 2021).

1.7.5 Dangerous roles

For military roles, where the reconnaissance of heavily defended areas is necessary, the attrition rate of a manned aircraft is likely to exceed that of a UAV. Due to its smaller size and greater stealth, the UAV is more difficult for an enemy air defence system to detect and more difficult to strike with anti-aircraft or missiles. Also, in such operations the concentration of aircrew upon the task may be compromised by the threat of attack. Loss of the asset is damaging, but equally damaging is the loss of trained aircrew and the political ramifications of capture and subsequent propaganda, as seen in the recent conflicts. The UAV operators are under no personal threat and can concentrate specifically, and therefore more effectively, on the task in hand. The UAV therefore offers a greater probability of mission success without the risk of loss of aircrew resources.

Power-line inspection and forest fire control are examples of applications in the civilian field for which experience, sadly, has shown that manned aircraft crew can be in significant danger. UAV can carry out such tasks more readily and without risk to personnel. Operating in extreme weather conditions is often necessary in both the military and the civilian field. Operators may be reluctant to risk personnel, and the operation, though necessary, may not be carried out. Such reluctance is less likely to apply with a UAV (Calcara et al., 2022).

1.7.6 Covert roles

In both military and civilian policing operations, there are roles where it is imperative not to alert the enemy (other armed forces or criminals) to the fact that they have been detected. Again, the lower detectable signatures of the UAV (see Chapter 7) make this type of role more readily achievable. Also in this category is the covert surveillance which arguably infringes the airspace of foreign countries in an uneasy peacetime. It could be postulated that in examples such as the Gary Powers/U-2 aircraft affair of 1960, loss of an aircraft to enemy territory could generate less diplomatic embarrassment if no aircrew were involved (Enemark, 2021).

1.7.7 Research roles

UAVs are being used in research and development work in the aeronautical field (Demir, 2015). For test purposes, the use of UAVs as small-scale replicas of projected civil or military designs of manned aircraft enables airborne testing to be carried out, under realistic conditions, more cheaply and with less hazard. Testing subsequent modifications can also be effected more cheaply and more quickly than for a larger manned aircraft, and without any need for changes to aircrew accommodation or operation. Novel configurations may be used to advantage for the UAV. These configurations may not be suitable for containing an aircrew (Ekka, 2017).

1.7.8 Environmentally critical roles

This aspect relates predominantly to civilian roles. A UAV will usually cause less environmental disturbance or pollution than a manned aircraft pursuing the same task (Ene et al., 2020). It will usually be smaller, of lower mass, and consume less power, thus producing lower levels of emission and noise. Typical of these are the regular inspection of power lines where local inhabitants may object to the noise produced and where farm animals may suffer disturbance both from noise and from sighting the low-flying aircraft (Jiménez López and Mulero-Pázmány, 2019).

1.7.9 Economic reasons

Typically, the UAV is smaller than a manned aircraft used in the same role, and is usually considerably cheaper in cost. Operating costs are less since maintenance costs, fuel costs, and hangarage costs are all less. The labour costs of operators are usually lower and insurance may be cheaper, though this is dependent upon individual circumstances (Grlj et al., 2022).

1.8 APPLICATIONS OF UAS (UNMANNED AERIAL SYSTEMS)

UASs are employed in many niche applications (Kardasz et al., 2016). The plethora of applications can be categorized into two distinct domains, namely, civilian domain and military domain applications. In the military, UASs are deployed where manned operations are risky, require persistence, and are often routine in nature. These operations are often known as D3 operations. UASs are deployed to perform distinct operations of land forces, air forces, and marine forces separately and often in a networked manner.

Often, discussing and categorizing the applications of UAS is quite motivating for first-time readers to start with. Therefore, before looking into the details of UAS, it is more relevant to discusses some of its applications in these domains and enlist them. UAS applications are emerging and are numerous, with the most obvious applications in distinct categories presented in the list below:

1.8.1 Civilian applications

a) Aerial photography, film, video, and still photography (Colomina, 2014).
a) Agriculture crop monitoring and spraying of pesticides (Restas et al., 2021).
b) Traffic monitoring and driving anomaly detection
c) Coastline and sea-lane monitoring for search and rescue operations
d) Customs and excise surveillance for illegal imports
a) Electricity companies' power-line inspection
b) Fire services and forestry fire detection, incident control

1.8.2 Military applications

a) Employing surveillance and ISTAR applications
b) Mapping and locating objects on battlefield
c) Providing a platform for extended communication line of sight (LoS)
d) Countering drone attacks on enemy objectives
e) Providing defence against enemy intrusion
f) Placing gun on drone (mounting weapon on drone)

1.9 CLASSIFICATION OF UNMANNED SYSTEMS

Since their inception, UASs have diversified in various forms, and different application-specific UASs have been designed and developed (Figure 1.5).

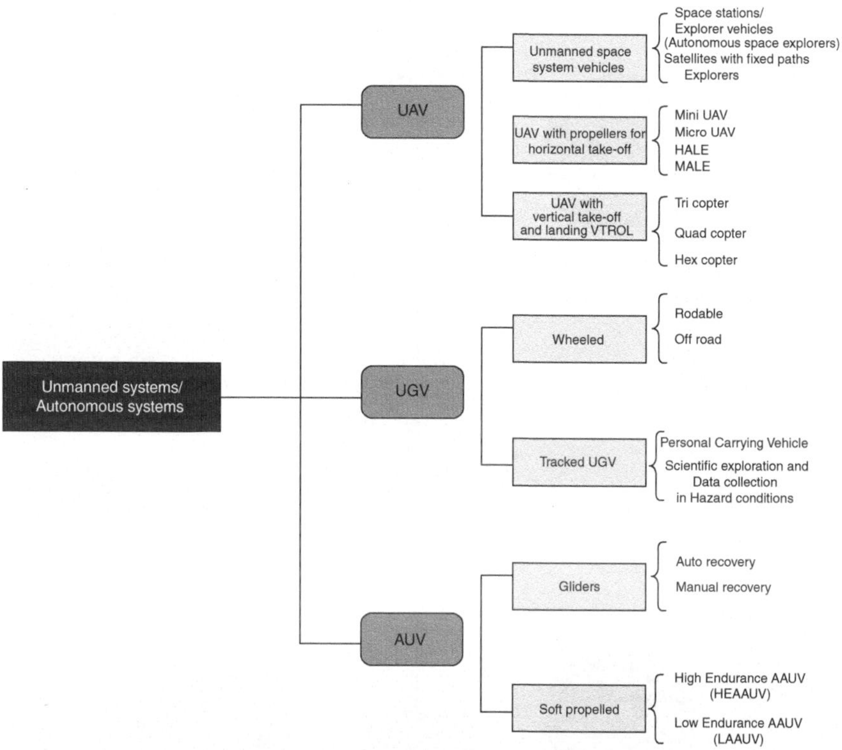

Figure 1.5 Classification of autonomous systems as per their motion and kinematic.

Different UASs are designed and deployed in different spatial domains, such as the Earth's surface, near air space, deep space, the sea surface, and deep under water for various purposes. The sensors deployed in these UASs vary in their sensing mechanism and operate in different bandwidth of the EM (Electro Magnetic) spectrum.

1.10 ORGANIZATION OF THE BOOK

This book comprises 10 chapters explaining the design, integration, and application of various unmanned systems. Out of the 10 chapters, the first seven are about UAVs; two chapters, Chapters 8 and 9, discuss Autonomous Underwater Vehicles (AUVs); and Chapters 10 and 11 are about Unmanned Ground Vehicles (UGVs). Chapter 12 examines the general concepts of perception, mapping, and navigation which are common among UAV, UGV, and AUV. In each of the chapters, the system view, communication view, and sensor view of the respective unmanned systems are discussed in detail. The specific design issues are considered with respect to the unmanned system. To make the concepts clear, the

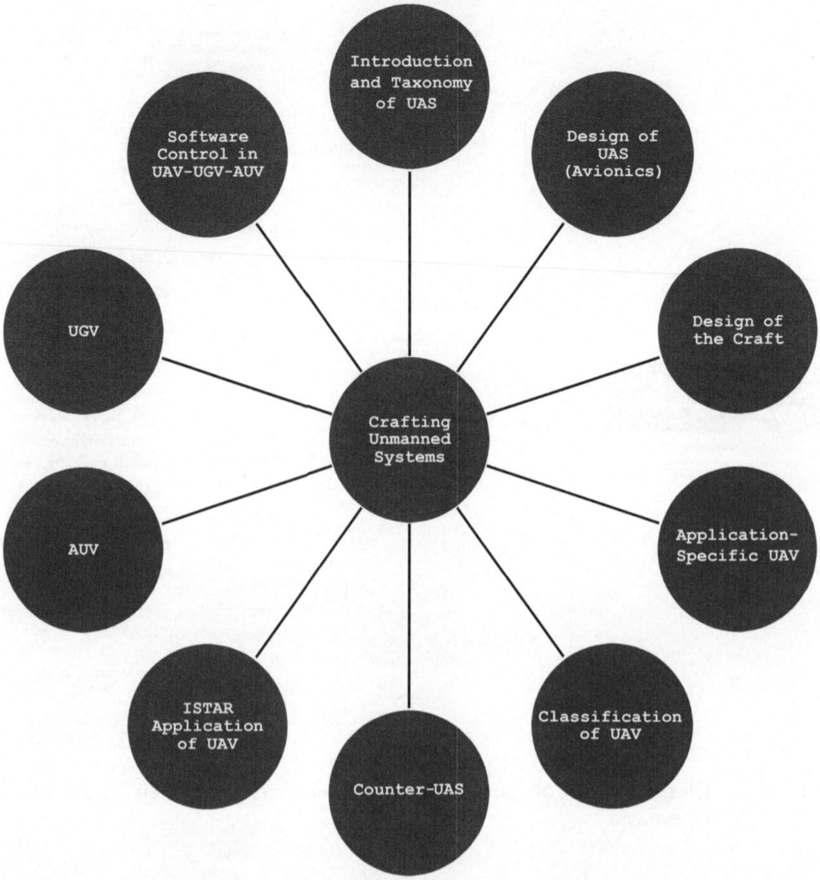

Figure 1.6 Various chapters and organization of the book.

application of the systems in various domains and scenarios are discussed. The chapters are full of wireframe models, real and existing systems, and tables containing various data and specifications of the UAS. The collaborative diagram (Figure 1.6) depicts a bird's-eye view of the 12 chapters of this book.

QUESTIONS AND ANSWERS

1. What is a UAS? What is a UAV? What is a drone?

Answer:

- UAS stands for unmanned aerial system, which is a system of systems consisting of a set of well-coordinated subsystems comprising the ground pilot or ground control system, the telemetry subsystem, the mobile data storage and analysis centre, the sensor payload, and the navigation system mounted on the UAV. A UAS is the totality of everything that makes a UAV, including its GPS module, ground control module, transmission systems, sensor, all the software, and the ground pilot or the person on the ground controlling the UAV.

- UAV stands for unmanned aerial vehicle, which is a specific type of aerial vehicle from the generic class of systems known as UASs. UAVs are aircraft that can fly without an on-board pilot and are expected to have autonomous flight capabilities.
- Drone is a term that is commonly used to refer to any UAV, regardless of its type or specific capabilities. In many ways, "drone" is a non-technical catch-all term that can be used for both UAV and UAS.

2. What is the difference between a drone and a UAV?

Answer: The difference between a drone and a UAV is that a drone is a term that can be used to refer to any unmanned aerial vehicle, regardless of its type or specific capabilities, while a UAV is a specific type of aerial vehicle that is part of a UAS and is expected to have autonomous flight capabilities. In other words, all UAVs can be called drones, but not all drones can be called UAVs. Additionally, a drone can refer to vehicles that can be controlled remotely, whereas UAVs are typically expected to have autonomous flight capabilities.

3. What are different components of a UAS? Explain each of the components with their functions.

Answer: The main components of a UAS are the UAV, the GCS, and the communication link between them. The UAV includes the airframe, propulsion system, avionics, and payload. The GCS includes the user interface, the control system, and the communication system. The communication link between the UAV and the GCS is used to transmit commands and telemetry data.

4. Give a broad classification of UAS and explain how UAS and the class of UAV differ from each other.

Answer: UAS can be classified based on their size, range, endurance, altitude, and mission type. UAVs can be classified into different classes based on their maximum take-off weight, which determines their regulatory requirements and operational capabilities.

5. List different types of sensors and payloads used by different types of UASs.

Answer: Different types of sensors and payloads used by UASs include cameras, multispectral sensors, LiDAR, radar, sonar, magnetometers, and various other types of environmental sensors. The specific sensor or payload used depends on the mission requirements of the UAS.

6. List different roles a UAS is employed for.

Answer: UASs are employed for various roles, including surveillance, reconnaissance, target acquisition, search and rescue, disaster response, scientific research, and commercial applications such as aerial photography and package delivery.

7. List 10 different military and 10 different societal applications of UAS.

Answer: Military applications of UAS include Intelligence, Surveillance, and Reconnaissance (ISR), target acquisition and tracking, combat operations, and logistics. Societal applications of UAS include law enforcement, border patrol, search and rescue, disaster response, environmental monitoring, and agricultural and infrastructure inspections.

8. What is a GCS? What are the different functions a GCS performs?

Answer: A Ground Control Station (GCS) is a system used to control the flight of an unmanned aircraft. It includes the user interface, the control system, and the communication system. The GCS performs functions such as monitoring the status of the UAV, sending commands to the UAV, and receiving telemetry data from the UAV.

9. Who is a ground pilot and what are their functions in a UAS mission?

Answer: A ground pilot is a person who operates the GCS and controls the flight of the UAV during a mission. Their functions include monitoring the status of the UAV, navigating the UAV to the target area, operating the payload systems, and ensuring safe operation of the UAV.

10. Give a classification of UASs from the perspective of its endurance of flight.

Answer: UASs can be classified based on their endurance of flight into three categories: short endurance, medium endurance, and long endurance. Short endurance UASs can fly for a few hours, medium endurance UASs can fly for several hours, and long endurance UASs can fly for more than 24 hours.

11. What is navigation? Give a brief description of different navigation systems used in UAS.

Answer: Navigation is the process of determining the position, velocity, and orientation of a vehicle relative to a reference frame. Different navigation systems used in UASs include GPS, inertial navigation systems, and visual odometry.

12. Why are UAS systems deployed and what are their advantages vis-à-vis manned systems?

Answer: UASs are deployed for a variety of reasons, including their ability to operate in hazardous environments, their cost-effectiveness compared to manned systems, and their ability to provide persistent surveillance and reconnaissance. UASs also have the advantage of being able to operate in areas where manned aircraft cannot, and they can be used for a variety of commercial applications.

REFERENCES

Calcara A., Gilli A., Gilli M., Marchetti R., Zaccagnini I. Why Drones Have Not Revolutionized War: The Enduring Hider-Finder Competition in Air Warfare. *International Security*. 2022; 46(4):130–171. www.muse.jhu.edu/article/855434

Colomina I., Mohna P. Unmanned Aerial System for Photogrammetry and Remote Sensing: A Review. *ISPRS Journal of Photogrammetry and Remote Sensing*. 2014; 92:79–97.

Demir K., Cicibaş H., Arica N. Unmanned Aerial Vehicle Domain: Areas of Research. *Defence Science Journal*. 2015; 65:319–329. 10.14429/dsj.65.8631

Ekka P. A Review Paper on Unmanned Aerial Vehicle (U.A.V.). *International Journal of Engineering Research & Technology (IJERT)*. 2017; 5(23): 1–3.

Ene M., Jianu R. Unmanned Aerial Systems for Environmental Monitoring: A Review. *Journal of Environmental Management*. 2020; 262:110323. 10.1016/j.jenvman.2020.110323

Enemark C. Armed Drones and Ethical Policing: Risk, Perception, and the Tele-Present Officer. *Crim Justice Ethics*. 2021 Jun 19;40(2):124–144. 10.1080/0731129X.2021.1943844. PMID: 34413570; PMCID: PMC8367046.

Franta N.R. Swarm Intelligence for Autonomous UAV Control. Naval Post Graduate School, Monterey, California; 2005 June.

Grlj C.G., Krznar N., Pranjić M. A Decade of UAV Docking Stations: A Brief Overview of Mobile and Fixed Landing Platforms. *Drones*. 2022; 6(1):17.

Hildmann H. et al. Nature-Inspired Drone Swamming for real time Data Collection under Dynamic Operational Constraints. *MDPI*. 2019 Sept.

Jimenez Lopez J., Mulero-Pazmany M. Drones for Conservation in Protected Areas: Present and Future. *Drones*. 2019; 3(1):10. https://doi.org/10.3390/drones3010010

Kardasz P., Doskocz J. Drones and Possibilities of Their Using. *Journal of Civil & Environmental Engineering*. 2016; 6. 10.4172/2165-784X.1000233

Kunhoth J., Karkar A., Al-Maadeed S. et al. Indoor Positioning and Wayfinding Systems: A Survey. *Human-centric Computing and Information Sciences*. 2020; 10:18. https://doi.org/10.1186/s13673-020-00222-0

Mohsan S.A.H, Khan M.A., Noor F., Ullah I., Alsharif M.H. Towards the Unmanned Aerial Vehicles (UAVs): A Comprehensive Review. *Drones*. 2022; 6(6):147. https://doi.org/10.3390/drones6060147

Restas A., Szalkai I., Gyula Ó. Drone Application for Spraying Disinfection Liquid Fighting against the COVID-19 Pandemic—Examining Drone-Related Parameters Influencing Effectiveness. *Drones*. 2021; 5:58. 10.3390/drones5030058

Valavanis K.P., Vachtsevanos G.J. (eds). Handbook of Unmanned Aerial Vehicles. Springer Science + Business Media, Dordrecht; 2015.

Vergouw B., Nagel H., Bondt G., Custers B. Drone Technology: Types, Payloads, Applications, Frequency Spectrum Issues and Future Developments. Chapter-2, 22–43, 2016; 10.1007/978-94-6265-132-6_2

Wu J., Yan F., Lu X., Wang J. Autonomous Target Detection and Tracking Using Unmanned Aerial Vehicles: A Review. *Journal of Intelligent & Robotic Systems*. 2017; 88(1):21–39. 10.1007/s10846-017-0647-y

Yaacoub J.P., Noura H., Salman O., Chehab A. Security Analysis of Drones Systems: Attacks, Limitations, and Recommendations. *Internet of Things*. 2020 Sep;11:100218. doi: 10.1016/j.iot.2020.100218. Epub 2020 May 8. PMCID: PMC7206421.

Zhang Y., Chen, L., Yang L. A Review of Visual Tracking for Unmanned Aerial Vehicles. *Journal of Intelligent & Robotic Systems*. 2019; 93(3–4):581–604. 10.1007/s10846-018-0996-4

Chapter 2

Crafting an Unmanned Aerial Vehicle (UAV)

OVERVIEW

This chapter discusses various design aspects of a basic unmanned aerial vehicle (UAV) (Sadraey, 2020). It starts with a knowledge graph depicting various subsystems and their functional components that are used in designing a generic UAV. Various alternatives available for each of the subsystems are discussed. A basic structure of a UAV is mechatronic structure with suitable avionic integrated into it. The propeller, which is the mechanism for the take-off, flying, and landing of a UAV is discussed. Also, a UAV has a structure for holding and controlling the sensors for navigation and payload sensors for sensing the environment. Each UAV is designed for a specific mission. The mission objectives guide the crafting and designing of the UAV with a specific structure, avionic, propulsion system, and payload sensors among the many available choices.

After choosing appropriate subsystems, the design process of a UAV follows with the process of integration of the subsystems (Sadraey, 2020). Often, the UAV starts with a simulation of the moving geometry of the structure of the craft with its sensor payload and other accessories. The simulation is performed to ascertain the aerodynamic stability of the platform in different aerodynamic conditions to meet the mission requirements. The endurance, range, and other mission objectives are ascertained through various laboratory simulations so that failure is minimized during the actual testing of the UAV.

2.1 INTRODUCTION TO THE UNMANNED AERIAL VEHICLE (UAV)

2.1.1 The system basis of the UAV

The Unmanned Aerial Vehicle (UAV) system comprises a number of elements, or subsystems, of which the aircraft is one. The technical functional structure of a typical system is shown in Figure 2.1. There are, of course, other integration facets when designing the more global system, such as the clearance required to operate within controlled airspace, which are not shown in this figure. These

DOI: 10.1201/9781003432685-2

issues are addressed in later chapters. It is always most important to view each subsystem of the UAV system as an integral part of that system. No one subsystem is more important than another, though some, usually the aircraft, have a greater impact upon the design of the other subsystems in the system than do others. In the early days of UAV development, for example, some unmanned aircraft were designed, and made, with inadequate regard to how payloads would be mounted, the aircraft launched or recovered, communications affected, or the system maintained and transported in conditions under which the system was required to perform. Subsequent attempts to build an operational system around it were either doomed to failure or resulted in unacceptable compromises or unacceptable cost.

Unmanned Aircraft Systems (UASs) are also usually, but not always, the centre in which the UAV mission is pre-planned, in which case it may be known as the Mission Planning and Control Station (MPCS). Less usually, the mission may be planned from a central command centre, and the mission data is sent to the Control Station (CS) for its execution. From the CS, the operators speak to the aircraft via the communications system uplink in order to direct its flight profile and to operate the various types of mission payload that it carries. Similarly, via the communications downlink, the aircraft returns information and images to the operators. The information may include data from the payloads, status information on the aircraft subsystems (housekeeping data), and position information. The launch and recovery of the aircraft may be controlled from the main CS or from a satellite (subsidiary) CS. The CS will usually also house the systems for communication with other external systems. These may include means of acquiring weather data, transfer of information from and to other systems in the network, tasking from higher authority, and the reporting of information back to that or other authorities.

2.1.2 The payload

The type and performance of the payloads is driven by the needs of the operational task. These can range from: (a) relatively simple subsystems consisting of an unsterilized video camera with a fixed lens having a mass as little as 200 g, through (b) a video system with a greater range capability, employing a longer focal length lens with zoom facility, gyro-stabilized, and with pan and tilt function with a mass of probably 3.4 kg, to (c) a high-power radar having a mass, with its power supplies, of possibly up to 1,000 kg. Some, more sophisticated, UAVs carry a combination of different types of sensors, within a payload module or within a series of modules. The data from these several sensors may be processed and integrated to provide enhanced information, or information which could not be obtained using a single type of sensor. For example, images from an optical (light) colour video camera, from a thermal (heat) imaging camera, and possibly from a radar scanner system may be fused together. Thus the thermal image and radar image may add information hidden to the optical image. The optical colour image will add discrimination, resolution, and contrast not available from the reduced contrast of the thermal image or the lower resolution of the radar image. Also, the reduction in performance of one sensor under differing light or atmospheric conditions of precipitation or pollution, may be compensated for by the complementary sensors. The images, or other data, obtained by these systems are processed into a form in which they can be transmitted via the downlink to the control station or other destination as appropriate. A number of different types of payloads appropriate for carriage by a UAV are described in Chapter 8.

2.1.3 The air vehicle

The type and performance of the air vehicle/aircraft is principally determined by the needs of the operational mission (see Chapter 5). The task of the aircraft is primarily to carry the mission payload to its point of application, but it also has to carry the subsystems necessary for it to operate. These subsystems include the communications link, stabilization and control equipment, power plant and fuel, electrical power supplies, and the basic airframe structure and mechanisms needed for the aircraft to be launched, to carry out its mission, and to be recovered. Other significant determinants in the design of the aircraft configuration are the operational range, air speed, and endurance demanded of it by the mission requirements (Hildmann, 2019). The endurance and range requirements will determine the fuel load to be carried. Achievement of a small fuel load and maximized performance will require an efficient propulsion system and optimum airframe aerodynamics.

2.2 BASIC STRUCTURE OF THE UNMANNED AERIAL VEHICLE

A UAV constitutes of four basic subsystems, namely, (a) the body of the craft along with its take-off and landing mechanism; (b) the payload sensors and navigation sensors; (c) the communication systems, and the command and control software; and (d) the Ground Control System (GCS) or ground pilot system (see Figure 2.1).

The above concept graph is transformed and actualized through a formal process of a design-assembly-testing-operation cycle. See Figure 2.2, which depicts a UAV with its functional components labelled. The design of the aerodynamic structure of the craft is specific to the mission needs and may vary for different UAVs.

The aerodynamic structure of the UAV is embedded with basic functional components such as its propeller, landing gear, avionic, flight control structures, radar, communication and data link to the GCS, gimbal to mount the payload

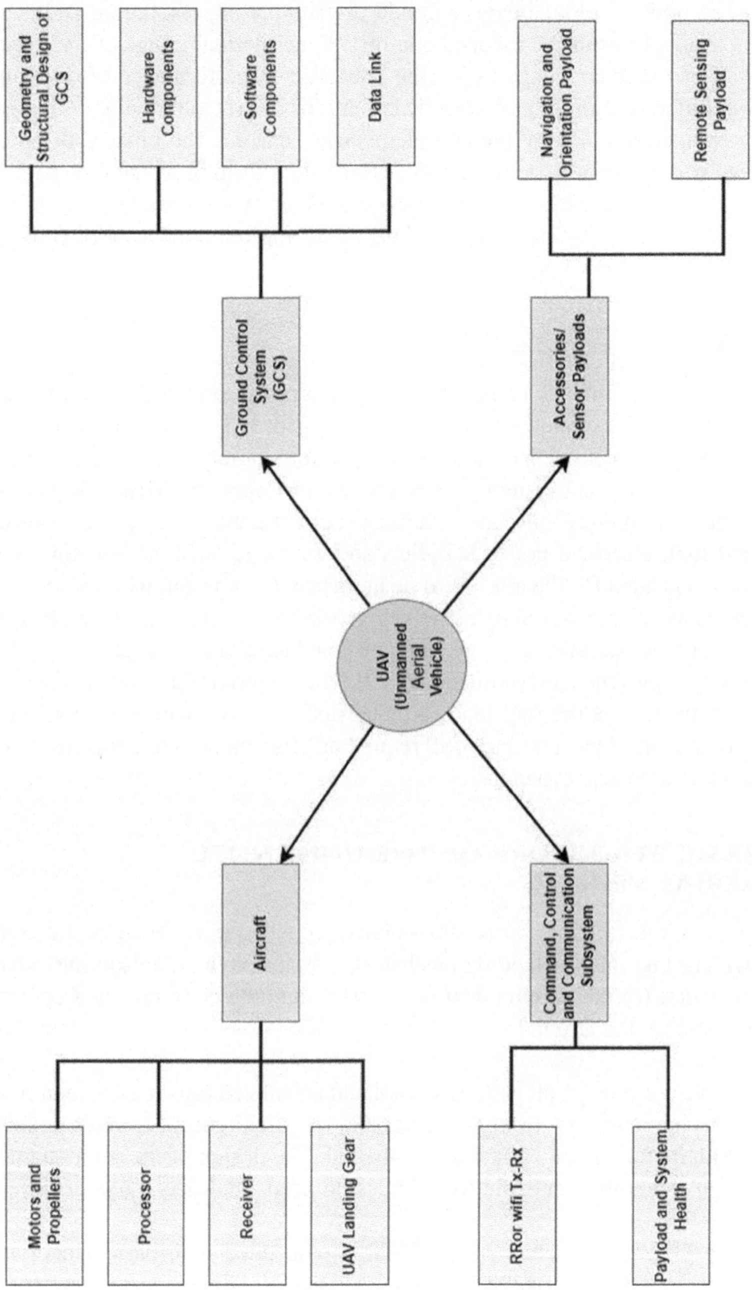

Figure 2.1 Knowledge graph of a basic UAV system with its subsystems.

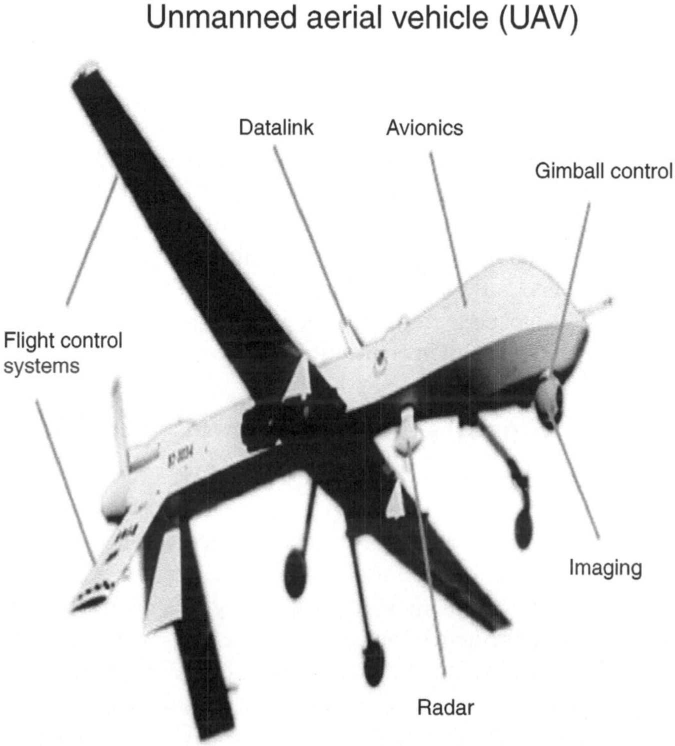

Figure 2.2 A conventional UAV with its functional components.

sensors (Barrows et al., 2003). In the next sections we discuss each of these subsystems and their alternates without any order or priority.

2.2.1 Ground control system

GCSs play a crucial role in the operation and control of UAVs. The type of GCS used can vary depending on the specific UAV and its mission requirements.

There are several variants of GCSs, each designed to cater to different needs. One common type is a small handheld remote controller, which provides a compact and portable interface for controlling the UAV. This type of GCS is typically used for small and lightweight UAVs, such as consumer drones or recreational models.

Another type of GCS is a simple laptop computer, which serves as the central command and control interface for the UAV. This allows operators to monitor and control various aspects of the UAV's flight, including its navigation, payload operations, and data collection.

For more complex missions or larger UAVs, a mobile vehicle may serve as the GCS. These vehicles are equipped with multifunctional displays and advanced communication systems to facilitate real-time data exchange and command link with the UAV. The GCS is typically installed within the vehicle, providing a dedicated and secure environment for mission control.

To maintain a continuous data and command link with the UAV, the GCS is equipped with antennae for transmitters and receivers (Tx-Rx). These antennae are usually mounted above the rooftop of the mobile enclosure, ensuring a reliable and constant communication connection between the GCS and the UAV.

2.2.2 Design of mobile handheld remote control

Depicted in Figure 2.3 is a wireframe model of one type of handheld remote controller which is mainly operated by a ground pilot for real-time control of the UAV in visual range. When the control stick is deflected, a potentiometer senses the deflection and gives a voltage signal to the microprocessor, the analogue signal is converted into digital signal corresponding to appropriate command, and is transmitted with the help of an antenna to the UAV with appropriate signalling protocol. The command is executed in the UAV for appropriate control of the UAV or its payloads. The UAV is equipped with a receiver antenna which sends a voltage signal to the on-board microprocessor. The microprocessor controls the servo attached to the control surface, thereby controlling the UAV's aerodynamic actions.

2.2.3 Structural design of a portable GCS

A portable GCS is similar to the simple handheld remote control in providing visual range and real-time control of the UAV. The additional feature available in a portable GCS is that it is capable of providing real-time video, image, and location, and UAV health monitoring data displayed in a digital map background with appropriate Graphical User Interface (GUI) and provides a graphical user interface.

2.2.4 Functional design of mobile GCS

A mobile GCS is a conglomeration of hardware, software, storage, and communication operating synchronously to perform the control and communication function of UAVs. The function of GCS can be executed through menu options available in GUI or through joystick control or through keyboard or mouse

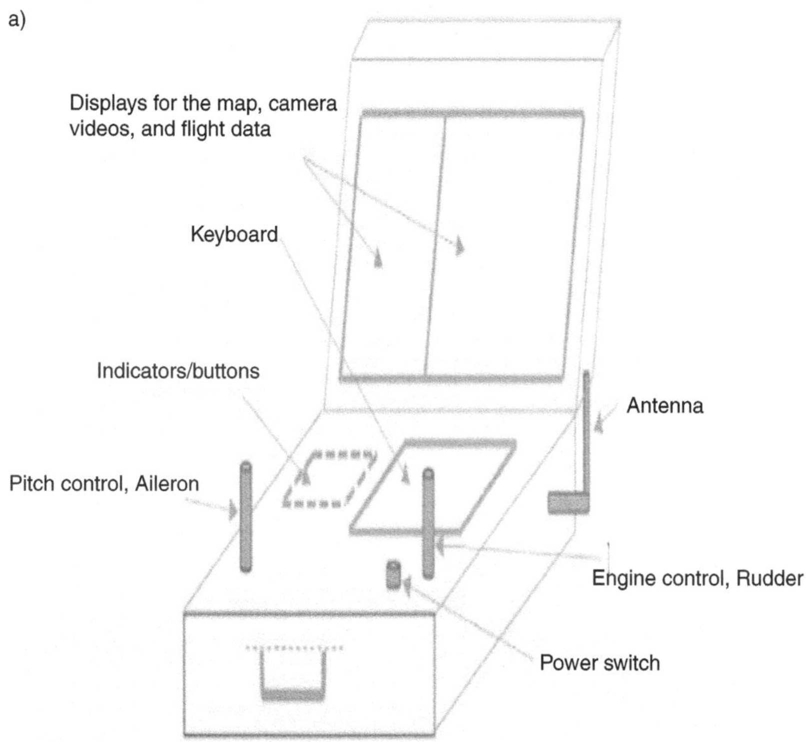

Figure 2.3 (a) wireframe model of portable remote control, (b) handheld remote controller. (Sadraey, 2020.)

control. The generic layout of a GCS housed in a mobile container is depicted in the block diagram in Figure 2.4.

The mobile GCS system is preferred for low-range medium-endurance UAVs. The multiple display systems provide various information like communication status, payload data, and navigation and control data. It is helpful to perform mission operations from an integrated environment. A mobile GCS makes the man-in-loop operations of the UAV more convenient and significantly decreases the cognitive load of the ground pilot. Also, this increases the operation and mission success to a high degree. Mobile GCSs are highly effective for on-sea operations and day and night operations, and therefore the CGS is mounted on the hull of ships.

b)

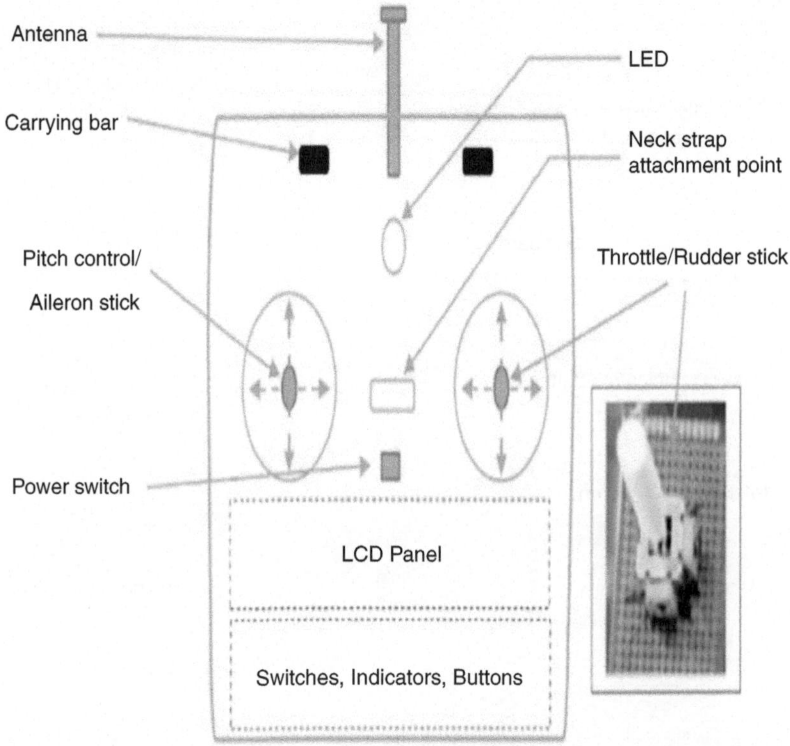

Figure 2.3 (Continued)

2.2.5 Hardware components of GCS

The GCS is often a ruggedized laptop or desktop with a capacity to operate in a high-density electromagnetically charged environment. Therefore, the system is ruggedized to operate with a certain level of tolerance to Electro Magnetic Interference (EMI) and have Electro Magnetic Compatibility (EMC) with other subsystems. Also, the system should be tolerant to the shock and vibration produced by the environment. It consists of the following hardware components

- Display screens: they are used to display the payload status, data, and communication status in addition to the attitude and position coordinates of the UAV on a digital map background.

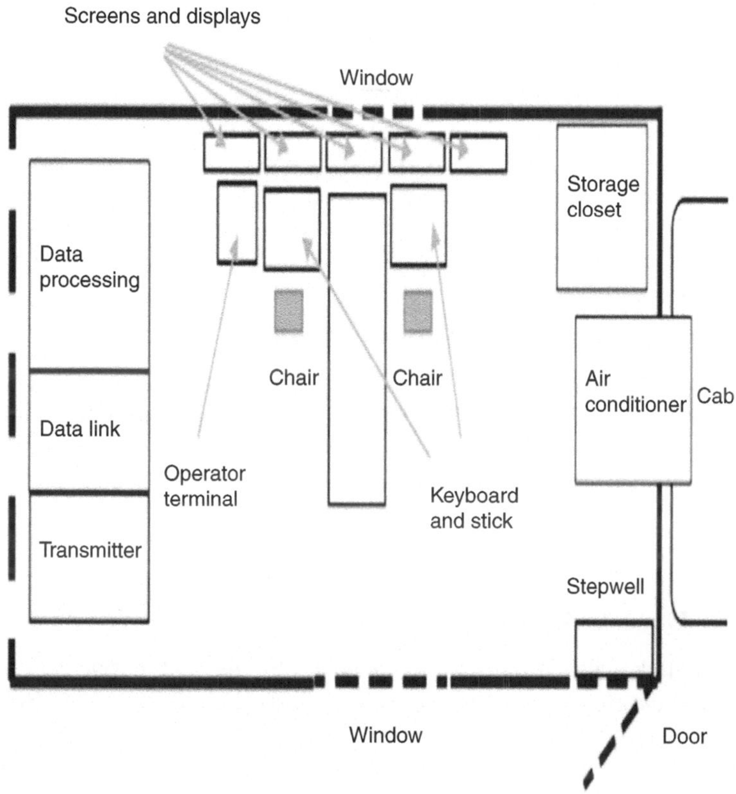

Figure 2.4 Mobile ground control system. (Mohammad H. Sadraey (2020), *Design of Unmanned Aerial Systems* (first ed., Wiley).)

- Control sticks (avionic control and payload control): they are used to deflect the control surfaces and engine throttle of the UAV. Generally, the payload control stick is used to control the operations of payloads like camera or ranging sensors. Generally, the control is to orient the focus of the sensor to a particular location of object in the field of view of the sensor. This operation is known as geo-pointing and geolocation of the payload sensor.
- Keypads: they are used to send the input mission data for the UAV from the GCS.

1. The data link has two components known as uplink and downlink. These communication links should be two non-interfering and

non-overlapping signal channels. The uplink channel is used to send commands and information from the GCS to the UAV (e.g. waypoints and command instructions) (Nils, 2013).

- Downlink is used to send the information from the UAV to the GCS (e.g. sensor data such as image frames, video or laser point cloud of the sensed space). The transmission channel consists of data terminal equipment and a wireless channel. A modem, controllers, and cryptographic equipment are the components of the data terminal equipment. The uplink and downlink channels are dynamically encrypted so as to avoid spoofing and hijacking of the UAV by unauthorized operators.

2.2.6 Software components of GCS

Software components of the GCS are what it make the GCS the powerful control centre of the UAV. The stack of software of GCS consists of the OS, the mission planner, the mission controller, and the post-mission data analyser. Software like the mission planner loads waypoints to the autopilot board to configure missions for the UAV. Alternate mission plans based on the UAV log data are also configured and uploaded by the mission planner software.

2.2.7 Data link

The data link makes use of radio frequencies to send and receive data from a UAV, and the frequency of the signal is majorly dependent on the UAV type and the mission requirements. A simple data link structure is depicted in Figure 2.5.

2.3 NAVIGATION AND ORIENTATION SYSTEM

Navigation and orientation of a UAV is a continuous operation performed by the UAV in real time. Performing navigation automatically by the UAV through the data obtained by the navigation sensors is what makes a UAV autonomous. There are different types of systems used to obtain navigation and orientation data of a UAV. Some of the most common types of navigation system that are used in UAVs are depicted in the hierarchical diagram in Figure 2.7. The navigation of unmanned systems is achieved mainly through three methods, namely, (a) Global Navigation Satellite Systems (GNSS) such as the Global Positioning System (GPS), the Global Navigation Satellite System (GLONASS), and GALILIO; (b) Inertial Measurement Systems (IMS) or Inertial Navigation Units (INU) in a GNSS-denied environment; and (c) a hybrid method using both GNSS and Inertial Navigation Systems (INS) combined.

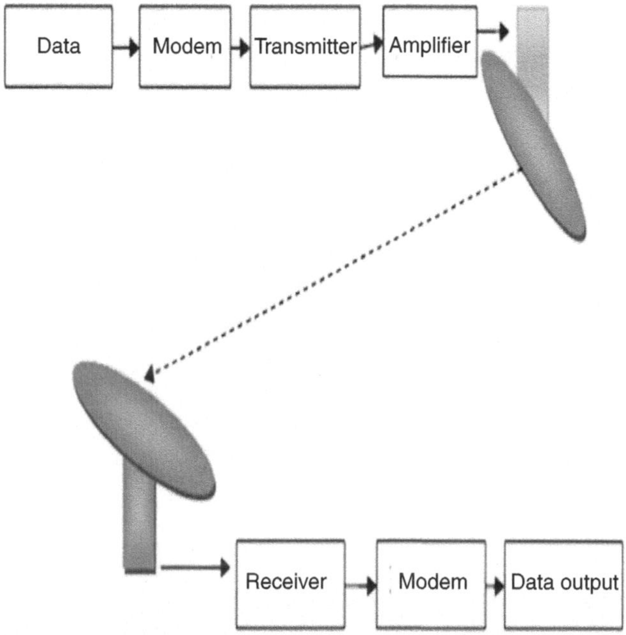

Figure 2.5 DATA link. (Mohammad H. Sadraey (2020), *Design of Unmanned aerial Systems* (first ed., Wiley).)

2.3.1 GNSS navigation system

The GNSS is based on a position calculation method on Earth's surface from the pseudo-distance data obtained from a minimum of three known satellite positions, and presence of a fourth satellite also aids in calculation of the altitude of the UAV.

The M-Code secure channel for GPS signals is making slow but steady progress towards becoming a reality for use in navigation. The highly encrypted M-Code to protect GPS signals from jamming and spoofing is currently enabled for use in secured communication between various UAVs and GCS.

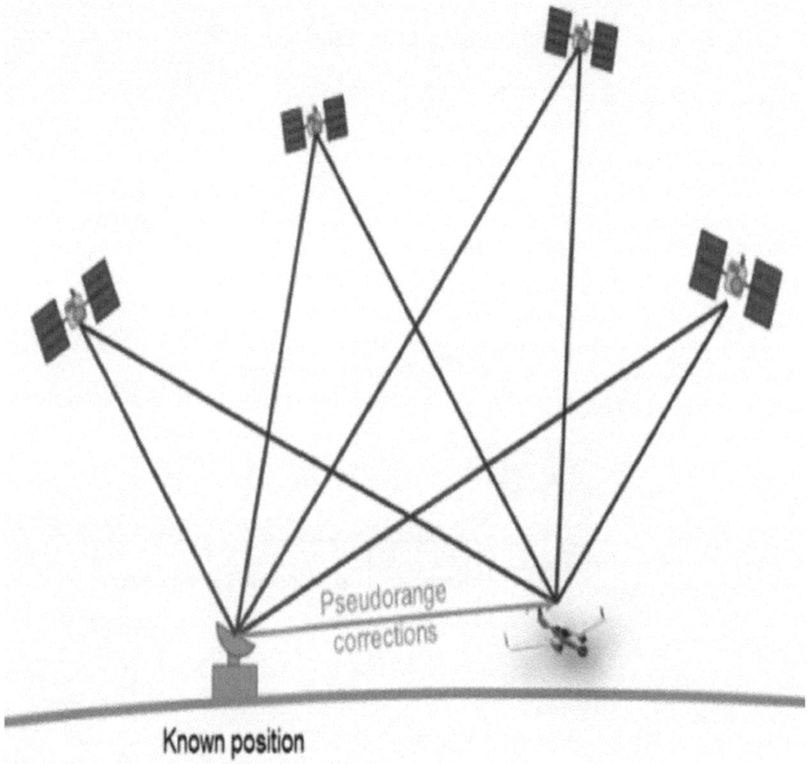

Figure 2.6 GNSS. (www.uavnavigation.com.)

2.3.2 Inertial measurement system

In an IMS there is the presence of an on-board computer that processes speed and altitude along with information provided by the sensors in order to obtain the current location from a predefined starting point. The IMS in turn has subsystems, as indicated in the following hierarchical diagram (Figure 2.8)

2.3.3 Inertial navigation unit

Inertial Labs has introduced the INS-BU, an advanced INS designed to cater to various platforms, including Unmanned Aerial Vehicles (UAVs), Unmanned Ground Vehicles (UGVs), and Unmanned Surface Vessels (USVs). The INS-BU incorporates a single-antenna U-Blox GNSS receiver, which enables 1 cm

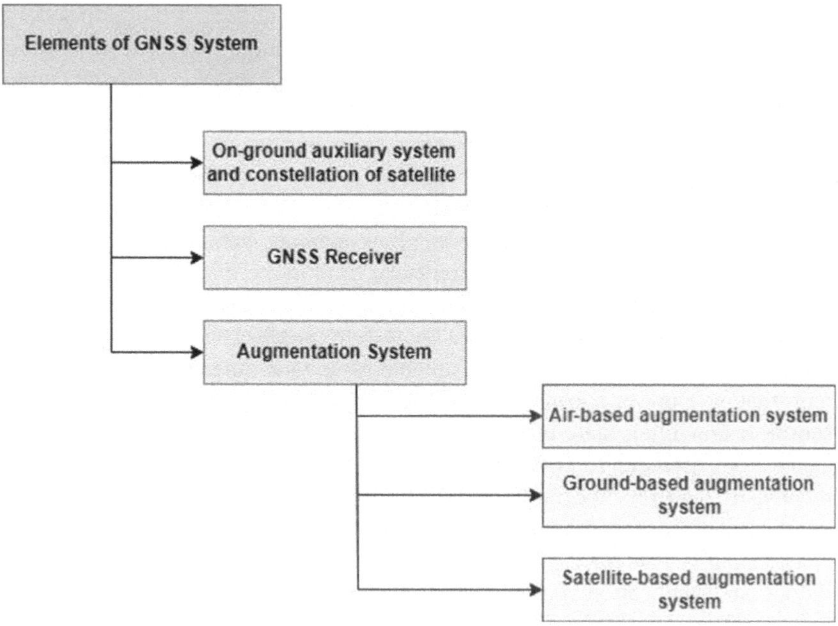

Figure 2.7 Most common types of navigation systems used in UAVs.

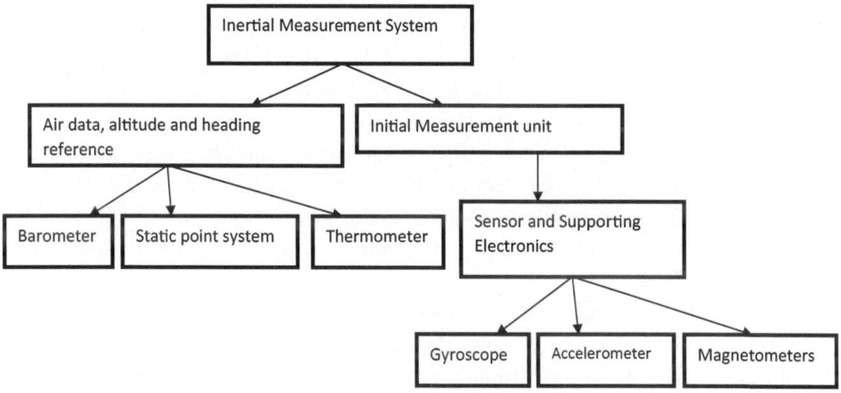

Figure 2.8 Elements of inertial measurement units (IMU).

RTK (real-time kinematic) positioning through RTCM3 RTK corrections. It supports all major GNSS constellations and frequencies, including GPS L1/L2, GLONASS L1/L2, Galileo E1/E5, BeiDou B1/B2, and QZSS L1/L2. This feature allows the INS-BU to establish the initial position of the UAV accurately.

The INS-BU offers a cost-effective solution and utilizes a range of aiding data for different applications. It excels in providing highly accurate navigation in GNSS-denied environments, thanks to Inertial Labs' enhanced Tunnel Guide Algorithm. The INS-BU integrates Inertial Labs' mini-AHRS (Attitude and Heading Reference System), which employs precision magnetometers, accelerometers, and gyroscopes along three axes to determine the device's orientation. The mini-AHRS leverages cutting-edge algorithms developed specifically for the motion of robots, unmanned and autonomous vehicles, and antennae. Inertial Labs' mini fluxgate magnetometers, used in the mini-AHRS, have had a long-standing reputation as reliable North reference sensors for over 70 years.

The INS-BU is highly versatile, with custom interfaces and low power consumption of only 2.5 W, making it suitable for a wide range of unmanned systems that have power consumption restrictions. It also incorporates an on-board sensor fusion filter, state-of-the-art navigation and guidance algorithms, and calibration software. To ensure secure positioning and navigation information, the INS-BU utilizes the on-board ZED-F9P receiver, which employs secure interfaces and advanced jamming and spoofing detection technologies.

The input signals for the INS-BU vary depending on the specific spatial conditions and application. For marine applications involving the navigation of unmanned surface ships or Unmanned Underwater Vehicles, the input signal is typically provided by a Doppler Velocity Log (DVL). In land applications for locating and navigating UGVs, inputs such as odometer, wheel sensors, encoders, and a Distance Measurement Instrument (DMI) are utilized. In aerial applications like UAVs, additional inputs such as wind sensors, airspeed sensors, and Doppler shift from a locator (used in long-term GPS-denied situations) are employed, along with an external Stand-Alone Magnetic Compass (SAMC/AHRS).

The output signals of the INS-BU encompass horizontal and vertical positions, pitch and roll angles, velocity, accelerations, angular rates, barometric pressure data, and Pulse Per Second (PPS) signals. The system also provides the data in the form of National Marine Electronics Association (NMEA) messages, including position, heading, pitch, and roll, which can be transmitted to the UAV and GCS.

A crucial characteristic of a navigation sensor, such as the INS-BU, is its Time To First Fix (TTFF), which should ideally be less than one second to facilitate highly agile navigation. Additionally, the rate at which it sends navigation data should be approximately 200 Hz, ensuring a quick start-up time and a data rate of around 200 updates per second.

2.4 NETWORK VIEW OF A UAV SYSTEM

A UAS can be conceptualized as a comprehensive network comprised of various systems and functional units that work together seamlessly. These include navigation sensors, payload sensors, data and communication links, GCSs, and the aircraft itself, all interconnected through a robust communication network. This network plays a crucial role in facilitating navigation, ground control operations, data transfer, payload communication, and other essential functions of the UAS (Mozaffari et al., 2018).

Given the high speed and altitude at which UAVs operate, it is impractical to rely on wired communication systems. Therefore, wireless communication systems play a vital role in enabling effective communication within the UAS network. The wireless communication system serves several crucial purposes, including the transmission of command and control signals between the UAV and the GCS. This ensures seamless coordination and control of the UAV's operations from the GCS (Geraci et al., 2018).

Furthermore, the wireless communication system facilitates the transfer of payload sensor data from the UAV to the GCS, enabling real-time monitoring and analysis of the collected data. This data transfer is vital for various applications such as remote sensing, aerial photography, surveillance, and environmental monitoring.

The wireless communication system also plays a critical role in flight path control, allowing operators to adjust and modify the UAV's trajectory and behaviour remotely. This capability enhances the flexibility and adaptability of the UAS for different mission requirements.

Additionally, the wireless communication system enables on-board data recording within the UAV and subsequent data transfer to the ground station. This feature ensures that valuable data collected during UAV missions can be safely stored and analysed post-flight, contributing to comprehensive data analysis and decision-making processes.

Overall, the wireless communication system within the UAS network is indispensable for efficient and reliable communication between the UAV and the GCS. It enables seamless coordination, data transfer, flight path control, and real-time monitoring, empowering operators to effectively utilize the capabilities of the UAS in a wide range of applications. A conceptual network of the UAS is depicted in the block diagram in Figure 2.9.

2.4.1 The network links of a UAV can be of the following types

(a) UAV to GCS link, which is a bidirectional link. This communication uses standardized industrial protocols; it can be a Bluetooth or Wi-Fi link when the UAV operates in near line of sight, and radio relay when it operates in a non-line of sight situation.
(b) UAV to GNSS or satellite, which is a unidirectional link from satellite to UAV receiving navigational information. This link from UAV to satellite is used for sending real-time navigation data with the help of GNSS or GPS

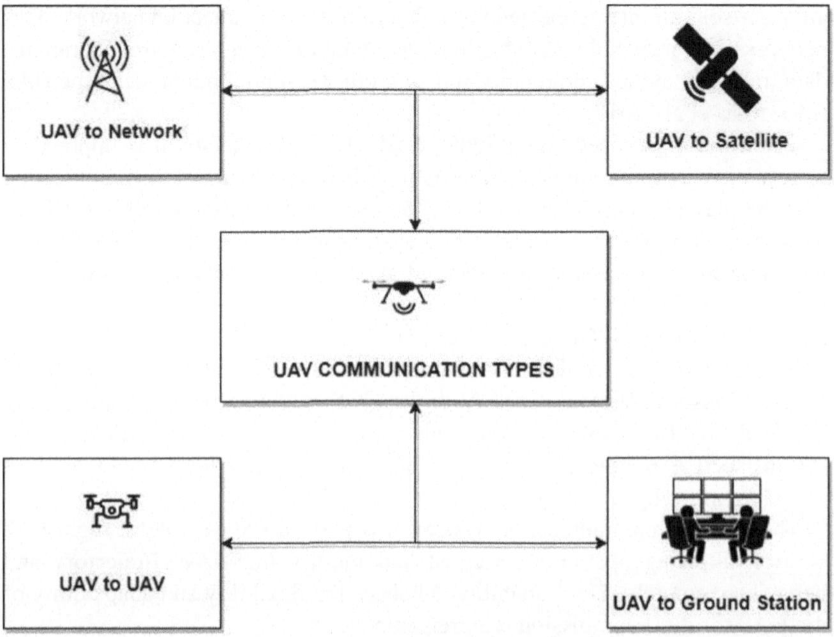

Figure 2.9 Possible communication links in a UAV system.

(c) UAV to UAV link, which is a link often used for a team of UAVs or a swarm of UAVs to function in a team (Natalie, 2005).

(d) UAV to fixed ground network, which includes cellular communication networks (2 GHz, 4 GHz, 5 GHz) (Li, 2019), helps to achieve a non-line of sight link between the GCS and the UAV.

Different modes and types of communications used in a UAS are depicted in a bubble diagram (Figure 2.10).

2.4.2 Short-range communication technologies used in UAVs

2.4.2.1 *Wi-Fi*

Wi-Fi is a type of short-range communication technology that is bound by a set of standards for designing of WLAN. It works on the radio bands of 2.4 GHz, 3.6 GHz, 5 GHz, and 60 GHz to help in the transmission of large-scale data like images and videos. The transmission range of a conventional Wi-Fi system is around 100m. Further extension of the range can be performed using ad hoc networking between the UAV. Some variants of the Institute of Electrical and

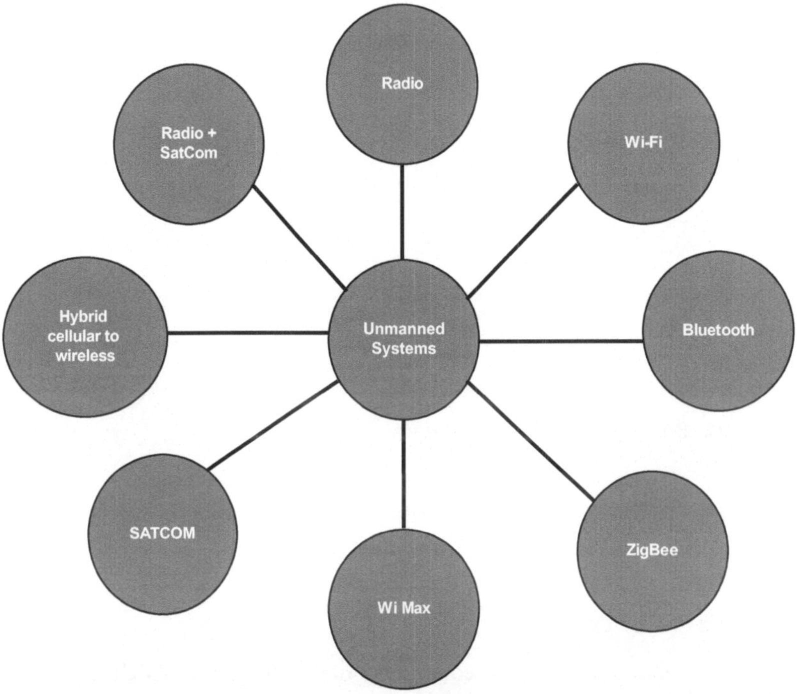

Figure 2.10 Different communication modes/protocols and media used in UASs.

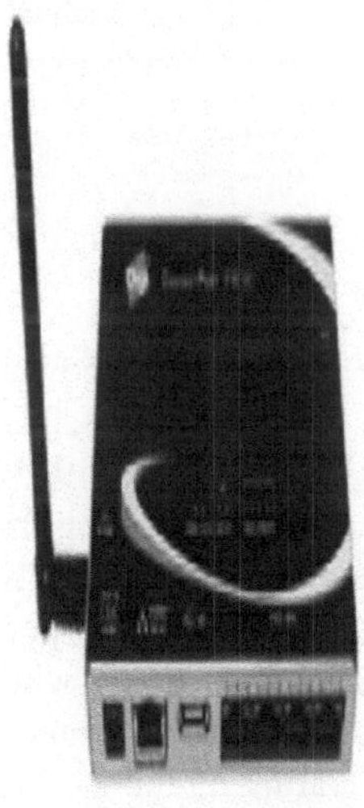

Figure 2.11 Wi-Fi device.

Figure 2.12 Bluetooth device.

Electronics Engineers (IEEE) standards like 802.11a/b/g/n/ac can be used for Flying Ad Hoc Network (FANET) applications.

2.4.2.2 *Bluetooth*

Bluetooth is also a short-range communication technology which operates in the frequency band of 2.4 GHz, and its communication range is around 10–200 m. There are multiple versions of Bluetooth based on their data rate; however, the maximum data rate that can be achieved is up to 24 Mbps. One of the latest Bluetooth versions is Bluetooth 5, which has features of improved speed, energy efficiency, transmission range, and coexistence along with other short-range communication systems.

2.4.2.3 *Zigbee*

ZigBee is an open, global, packet-based protocol designed to provide an easy-to-use architecture for secure, reliable, low power wireless networks. It operates in a short-range between 10 to 100 meters. Zigbee operates on unlicensed ISM bands, primarily 2.4 GHz (worldwide), 915 MHz (Americas/Australia), and 868 MHz (Europe), with data rates ranging from 20 kbps to 250 kbps. The operating range can be extendable with mesh networking. Zigbee devices are designed for low power consumption, often operating on batteries for years, and utilize "sleep mode" to conserve energy when inactive, with power consumption typically ranging from 0.1 to 0.5 watts. Zigbee uses a data rate ranging from 20 to 250 kbps, operating on the 2.4 GHz, 868 MHz, and 915 MHz frequency bands, making it suitable for low-power, short-range applications like sensor networks.

Figure 2.13 Zigbee device.

2.4.2.4 WiMAX

WiMAX technology is capable of providing high-quality voice and video streaming. It is most appropriate for UAV applications in rescue systems. It is a technology standard that delivers broadband access for long range in different ways, from point-to-point link to mobile cellular-type access. It has accommodation for both fixed and mobile broadband applications. The data rate for fixed application is 3 to 5 Mbps per use, and for mobile applications it is up to 30 Mbps.

2.4.3 Long-range communication technologies used in UAVs

2.4.3.1 Long-Term Evolution (LTE)

Long-Term Evolution (LTE) technology offers several advantages that make it well-suited for long-range communication applications in UAVs. With its secure wireless connectivity, high data rate, and mobility support, LTE provides a reliable and efficient communication solution.

One of the key benefits of LTE is its scalable bandwidth options. It can support various bandwidths, including 10 MHz, 15 MHz, and 20 MHz, and can even accommodate narrower bandwidths of less than 5 MHz. This flexibility allows for efficient spectrum allocation and utilization based on specific requirements and network conditions.

LTE supports both Frequency Division Duplexing (FDD) and Time Division Duplexing (TDD), enabling simultaneous two-way communication. This flexibility in duplexing methods ensures efficient use of available spectrum resources and facilitates bidirectional data transmission between the UAV and the GCS.

In terms of coverage, LTE has the capability to provide acceptable performance over long distances, with a cell size of up to 100 km. However, the optimum cell size for optimal performance is typically around 5 km. This makes LTE

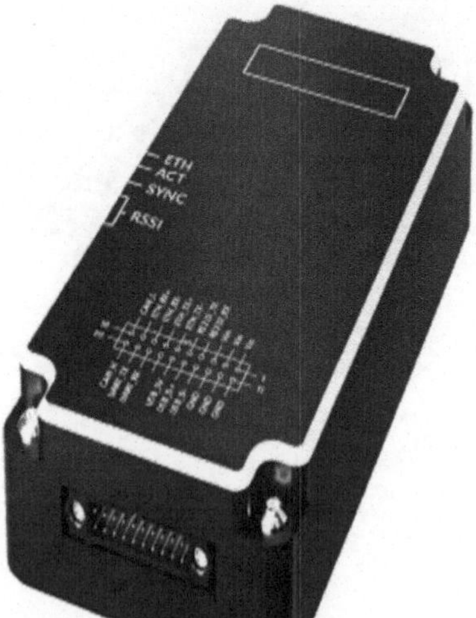

Figure 2.14 WiMAX device.

suitable for applications requiring communication over extended ranges, such as UAVs used in disaster management and surveillance scenarios.

The LTE network's ability to provide quality real-time video streaming and surveillance is particularly valuable in UAV operations. These capabilities enable seamless transmission of video feeds and surveillance data, ensuring timely and accurate information is delivered to the GCS. This is essential for effective decision-making and situational awareness during UAV missions.

LTE technology is especially suitable for low-altitude flights, where UAVs can take advantage of the LTE network's coverage and performance capabilities. By leveraging LTE's secure and high-speed connectivity, UAVs can maintain reliable communication links with the GCS, even in challenging environments.

2.4.3.2 Cellular mobile communication

In recent years, the evolution of mobile communication networks has witnessed significant advancements, from the introduction of 2G to the subsequent generations of 3G and 4G. However, the latest generation, 5G, holds great promise with its transformative features, including energy efficiency, reduced latency, high data rates, enhanced system capacity, and ubiquitous connectivity. Although 5G networks are still in the process of development and deployment, they are expected to be available in the near future, bringing remarkable improvements to communication technologies. (Song et al., 2019)

One of the key advantages of 5G networks, as highlighted by Bin Li (2019), is their exceptional speed, estimated to reach approximately 100 GB/s per user. This exponential increase in speed will revolutionize the way data is transmitted, enabling faster downloads, seamless streaming of high-resolution content, and efficient real-time communication. Moreover, the capacity of 5G networks is anticipated to be nearly 1,000 times greater than that of previous generations, allowing for a vast number of connected devices and applications without compromising network performance (Oubbati et al., 2020).

In the context of UAVs, 5G networks bring significant benefits. They can reduce the complexity and workload associated with connecting dynamic FANET systems to the main network. This simplification of connectivity enables smoother and more efficient transfer and streaming of live data from UAVs to GCSs or other end users. Additionally, backhauling requirements, which involve the transmission of data between remote UAVs and the central network, can be effectively addressed by leveraging micro base stations and establishing connections between the UAVs and the base band unit.

It is important to note that the connectivity and performance of 5G networks depend not only on the technological advancements but also on meticulous network planning and efficient service provision. Proper deployment and optimization of network infrastructure, along with the availability of 5G-compatible devices and applications, play crucial roles in realizing the full potential of 5G technology (Li et al., 2018).

In conclusion, the advent of 5G networks signifies a remarkable leap forward in mobile communication capabilities. With its energy-saving features, reduced latency, high data rates, enhanced system capacity, and ubiquitous connectivity, 5G is poised to revolutionize the way we communicate and interact with technology. The anticipated speed of 100 GB/s per user and the capacity increase of nearly 1,000 times will unlock new possibilities for UAVs and other connected devices, paving the way for innovative applications and services in various domains (Ullah et al., 2020).

2.4.3.3 *Satellite Communication (SATCOM)*

Satellite Communication, also known as SATCOM, plays a crucial role in facilitating beyond-visual-range communication for UAVs. It involves the transmission of electromagnetic waves between the GCS and the UAV, enabling long-distance communication and data transfer. Satellites act as relays, amplifying the signals between the UAV and the ground station.

SATCOM offers several advantages, particularly in situations where direct line-of-sight communication is not feasible. By utilizing satellites, UAVs can establish communication links over vast distances, extending their operational range and capabilities. This is particularly valuable for intelligence,

Table 2.1 Characteristics of different communication systems

Communication Technology	*IEEE Standard*	*Frequency/ medium*	*Spectrum Type*	*Device Mobility*	*Theoretical date rate*	*Range Indoor-outdoor*	*Network Topology*	*Latency*	*Advantage*	*Limitation*
Wi-Fi[7,8]	802.11	2.4 GHzIR	Unlicensed	Yes	Up to 2 Mbps	20m-100m	Ad hoc, star, mesh, hybrid	<5ms	high speed	limited range
	802.11a	5 GHz	Unlicensed	Yes	Up to 54 Mbps	35m–120m	Ad hoc, star, mesh, hybrid			
	802.11b	2.4 GHz	Unlicensed	Yes	Up to 11 Mbps	35m–140m	Ad hoc, star, mesh, hybrid			
	802.11n	2.4/5 GHz	Unlicensed	Yes	Up to 600 Mbps	70 m–250 m	Ad hoc, star, mesh, hybrid			
	802.11g	2.4 GHz	Unlicensed	Yes	Up to 54 Mbps	38 m–140 m	Ad hoc, star, mesh, hybrid			
	802.11ac	5 GHz	Unlicensed	Yes	Up to 3,466 Mbps	35 m–120 m	Ad hoc, star, mesh, hybrid			

Bluetooth 5[9-12]	802 15.1	2.4 GHz	Unlicensed	Yes	Up to 2 Mbps	40 m–200 m	Ad hoc, piconet	3ms	Energy efficient	Low data rate
Zigbee[13-15]	802 15.4	2.4 GHz	Unlicensed	Yes	250 kbps	10 m–100 m	Ad hoc, star, tree, cluster	15ms	Low cost	Low data rate
WiMAX[16-18]	802 16a	2 to 11GHz	Licensed	Yes	Up to 75 Mbps	Up to 48 km	Wide-area Wireless backhaul	30 ms	High throughput	Interference issues
LTE[19-22]	LTE	Up to 20 MHz	Licensed	Yes	Up to 300 Mbps	Up to 100 km	Flat, Ip based	5 ms	High bandwidth	Expensive
5G [23-29]	5G (eMBB)	28 GHz	Licensed	Yes	Up to 20 Gbps	Wide area	IP Based	1 ms	High data rate	Expensive
Satellite [30,31]	Satellite	Up to 40 GHz	Licensed	Yes	Up to 1 Gbps	World wide	-	550 ms	Wide coverage	High delay and high cost

Note: Credits: **https://doi.org/10.3390/drones3010016**

reconnaissance, and surveillance missions, where real-time data transmission is critical. SATCOM allows for the timely delivery of data obtained by UAVs in remote locations, eliminating the need to wait for the UAV to return to the base station.

Satellites used for UAV SATCOM can be either geostationary or Low-Earth Orbit (LEO). Geostationary satellites remain fixed at a specific point above the Earth's equator, providing continuous coverage over a large area. On the other hand, LEO satellites orbit the Earth at lower altitudes, which can help reduce signal latency but requires continuous tracking to maintain communication. The choice of satellite depends on specific mission requirements and system capabilities (Zhi et al., 2019).

Different frequency bands are allocated for satellite communication, each with its unique characteristics. Commonly used frequency bands include L, S, C, X, Ku, and Ka bands. The C band, for instance, utilizes a 6 GHz uplink band and a 4 GHz downlink band. Government and military applications often utilize the X band, operating with an 8 GHz uplink and a 7 GHz downlink. The Ka band, which operates at a higher frequency, is still in the development stage and uses a 30 GHz uplink and a 20 GHz downlink.

While satellite communication offers increased coverage and high-quality data transmission, it does have some limitations. Bandwidth availability can be a challenge, particularly in situations where multiple UAVs or other users compete for limited bandwidth resources. Additionally, the cost of data transmission via satellite communication can be relatively high, particularly for large volumes of data.

In summary, satellite communication plays a vital role in enabling beyond-visual-range communication for UAVs. It allows for the transmission of intelligence, reconnaissance, and surveillance data obtained by UAVs in remote locations, providing real-time information to the GCS. Despite limitations such as bandwidth constraints and cost considerations, satellite communication offers an invaluable solution for extending the operational capabilities of UAVs and enabling efficient data transfer over long distances.

2.4.4 Hybrid communication systems used for UAVs

2.4.4.1 Cellular and wireless hybrid

When using cellular network alone for UAV communication, the coverage and flight altitude are restricted. The major problems in cellular network is the signal interference leading to unreliable communication. The combination of Wi-Fi along with a cellular network widens the access network and enhances the network connectivity.

A simple module of integration of the mobile network along with Wi-Fi has been proposed in Sadraey (2020), where the UAV used an open-source flight controller which was integrated along with an advanced Reduced Instruction Set Computer (RISC) machines processor architecture unit; the UAV contained a single board computer along with a serial port connecting to the flight controller. The board computer had two major assignments: a gateway and an on-board GCS command centre. Through the USB port, a 4G/LTE dongle is connected to the on board computer. The dongle offers internet access to the GCS software installed in the board computer, thereby enabling the pilot to control the flight controller with the help of the Remote Desktop Protocol (RDP)/Secure Shell (SSH)/ Internet Protocol (Barrows et al., 2003). The communication between UAV and External Device (ED) is achieved by executing authentication between

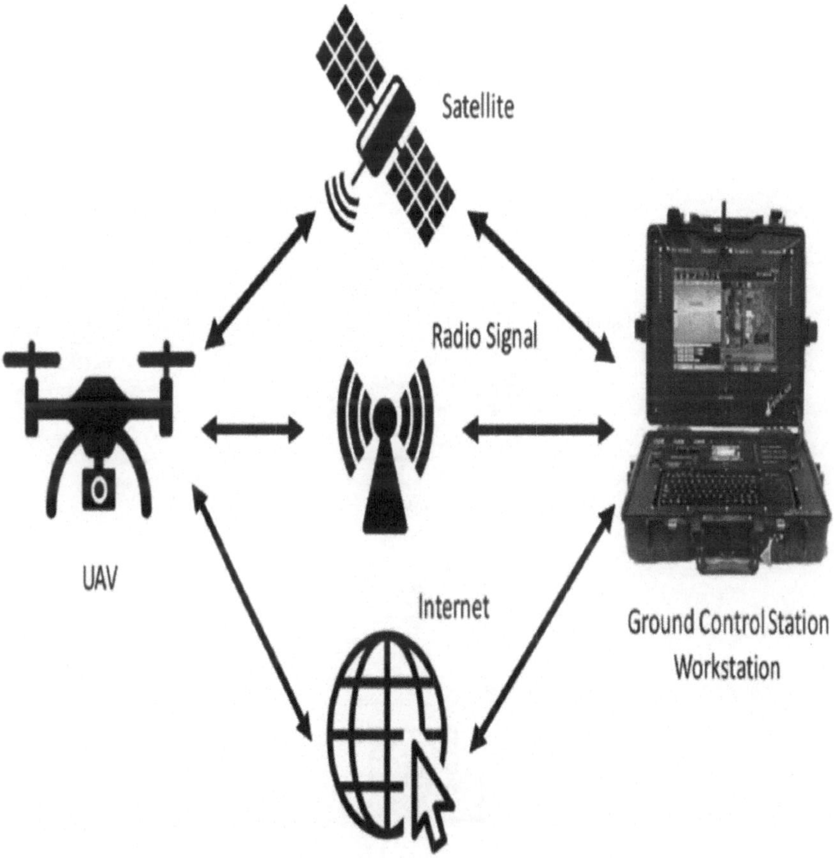

Figure 2.15 Cellular and wireless hybrid communication systems.

the ED and a single board computer; further, a secured tunnel is initialized directly to the UAV. The tunnel has encryption keys on both sides (ED and UAV). The ED can be a PC, mobile phone, or a tablet with an operating system that can connect the internet and run SSH or RDPs. Through the ED the pilot will be able to send a mission command, display, and store data securely in a created tunnel.

2.4.4.2 Bluetooth and cellular

In this hybrid communication, both the short-range technologies are used to obtain a system that has features such as low power consumption, low-cost, high range, and high speed. There are several variants of Bluetooth communication protocol present, namely Bluetooth master, Bluetooth slave, and

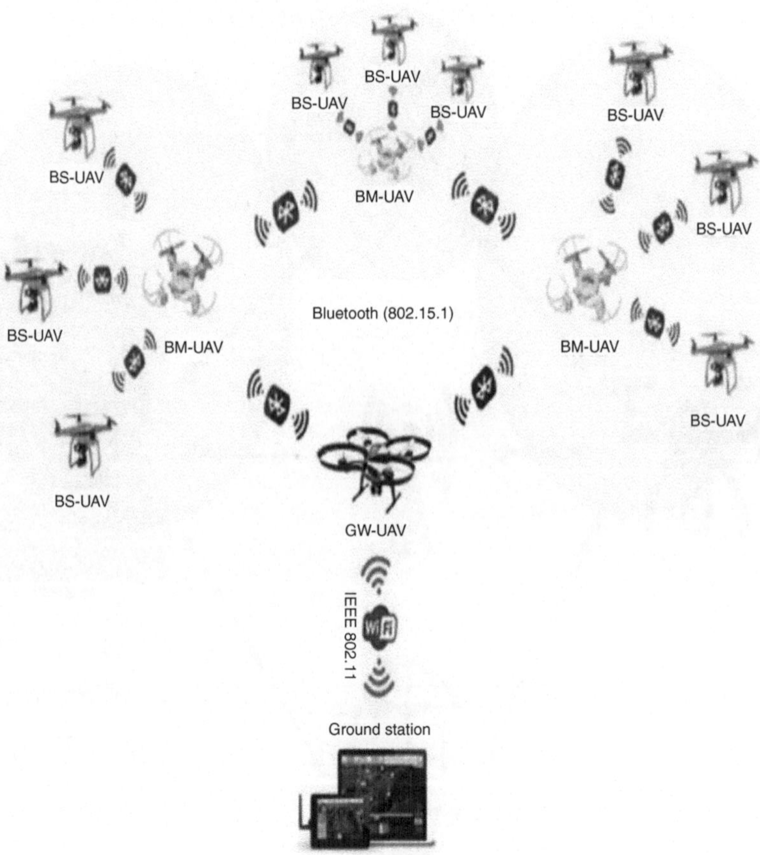

Figure 2.16 Sample topology for multilayered FANET. (https://doi.org/10.3390/drones3010016.)

Bluetooth gateway which are used judiciously with Wi-Fi to achieve UAV to GCS communication.

Initially, the connection between the UAVs is achieved using a beacon with the help of a discovery function. On successful communication between the UAVs, the data is transferred between the Bluetooth master-UAV and the Bluetooth slave-UAV using Bluetooth signals, and the communication between the Gateway-UAV and the GCS is achieved using Wi-Fi signals.

2.5 CHALLENGE OF UAV COMMUNICATION AND POSSIBLE SOLUTIONS

2.5.1 Multipath signal fading in large class of UAVs used for civilian purposes

The signal transmitted between the UAV and the GCS is required to be highly reliable and robust because otherwise it could pose a danger to other aircraft in flight or to humans on the ground. Some of the parameters used to check the performance of the communication system are the following:

- Availability: this is defined as the probability that an operational communication transaction can be initiated when needed.
- Integrity: refers to the probability that communication transactions are completed within the communication transaction time with undetected error.
- Continuity: Refers to the probability that an operational communication transaction can be completed within the communication transaction time.
- Latency and communication transaction time: latency is the maximum time taken for a one-way-message to be delivered whereas communication transaction time is the time taken for completion of the operational communication (Sadraey, 2020) transaction inclusion of the human response time.

2.6 DIFFERENT LANDING GEAR SYSTEMS OF UAVS

Landing gear is an important subsystem of any UAV. There are two different types of landing gear systems prevailing for the landing of UAVs. The first types of fixed landing gear are usually designed for multi-copters and are useful in the landing of lightweight UAVs (see Table 2.2). The second class of landing gears are retractable landing gear systems. These systems are designed to deploy at the time of landing of the UAV and get folded when the UAV is in operation (see Table 2.3).

Table 2.2 Some of the fixed landing gear designs used in multi-copter UAVs

 (a)	Simple vertical structure supports individual rotor arm at the rotor end
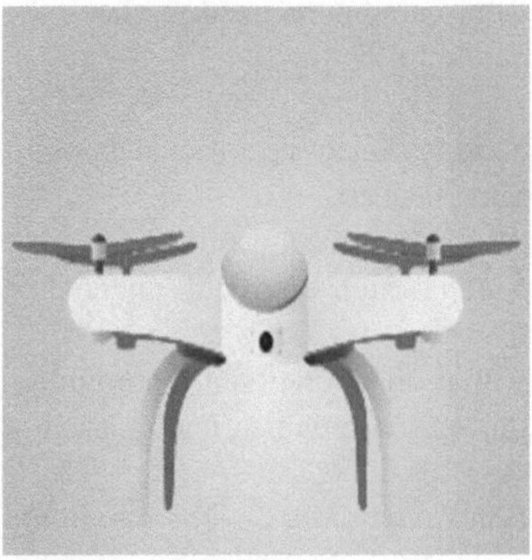 (b)	Four vertical structures connected to the centre of the chassis frame
 (c)	Vertical structure attached to the centre of the chassis with two vertical structures connected with a horizontal member

Table 2.2 (Continued)

 (d) (www.unmannedsystemstechnology.com//)	Two vertical members connecting one end to the centre of the chassis frame and the other end to horizontal skid-like members
 (e) (https://cfm-air.net/projects/drone-multicopter/)	Three vertical members connected to the centre of the chassis frame consisting of horizontal modifications for support at the other end

2.7 TESTING OF LANDING GEAR OF UAVS

Landing gear is a subsystem of the UAV that is responsible for withstanding the loads shock and vibration during take-off and landing, or during vertical drop scenarios. In the event of a crash or a hard landing, there is a very great possibility of landing gear touching the ground first, and if the structure is weak, this would lead to a major loss of the complete system. According to statistical data

Table 2.3 Some retractable landing gear designs

 (f)	Vertical members connected to the individual rotor arm folding to align with the rotor arm
 (g)	Individual landing structure folding outwards, aligning parallel to the rotor arm
 (h)	

Table 2.3 (Continued)

 (i) (https://drones.altigator.com/landing-gears-fixed-landing-gears-c-196_65_256.html?pg=store)	Two vertical structures interconnected and folding outwards to align horizontally

published by the US military, among all drone crashes from 2008 to 2019, 20% of crashes happened while landing. Hence, design and performance analysis of landing gear is a crucial step that needs to be performed for designing of UAV systems.

Standard impact tests are conducted on passenger aircraft and military aircraft. Companies such as Airbus and Boeing have their methods for drop testing. They make sure that the landing gear is capable to handle all exception and adverse landing cases before they fit it into the aircraft. However, there is very little research on impact tests on small drones. Impact test techniques are classified into two major types, experimental and computational, and they can be used for the validation of each other.

The experimental method involves drop tests of the model with accelerometers or a strain gauge, free fall of the body, hammer tests on the model, and so on. They have the disadvantage of expenditure and the possibility of instrumental error. Experimental set-up, instruments, and acquisition systems are necessary to acquire continuous data. The goal of an experiment is to measure the impact load acting on the model, and the deformation of the model. More information can be acquired based on the sensors and the acquisition system used. A few experimental techniques are shown in Figure 2.17.

Computational methods involve simulation and analysis of the landing gear model using different kinds of computational software. Several methods such as the finite element approach, virtual energy method, and so forth, are

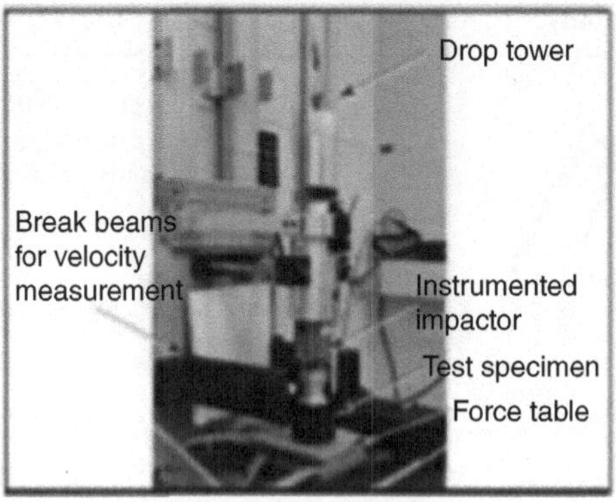

Figure 2.17 Guided drop test.

adapted for computational purposes. Software such as ANSYS, NASTRAN, and ABAQUES can be used to solve the model. They usually employ the finite element method for this purpose. A computer-aided design model of the required structure can be meshed and solved using a wide range of solvers according to requirements. The stress-strain distributions, deformation details, impact load details, and any other required data can be found based on the solver used. This has the limitations of computational requirements such as efficient systems, the need of experienced human resources to verify the models, and so on, and the advantages of saving money, changing the design before manufacturing, and having the opportunity to experiment on more parameters.

If the necessary computational method is obtained, then the various methods can be adapted to improve the impact performance of the landing gear based on the computational results (Liu et al, 2022). A few of the methods adapted to improve the performance and strengthen the landing gears of small drones are retractable landing gears, crash-proof structures around the vehicle, the introduction of struts in the landing gears, parachute systems, bio mimicking landing gears, and so forth. These are represented in Figure 2.18.

2.8 PRECISION LANDING OF UAVS

Having discussed and reviewed different landing mechanisms used in UAV landing, this section discusses smooth and precision landing of UAVs and answers some questions such as "What is precision landing?", "Why is precision landing

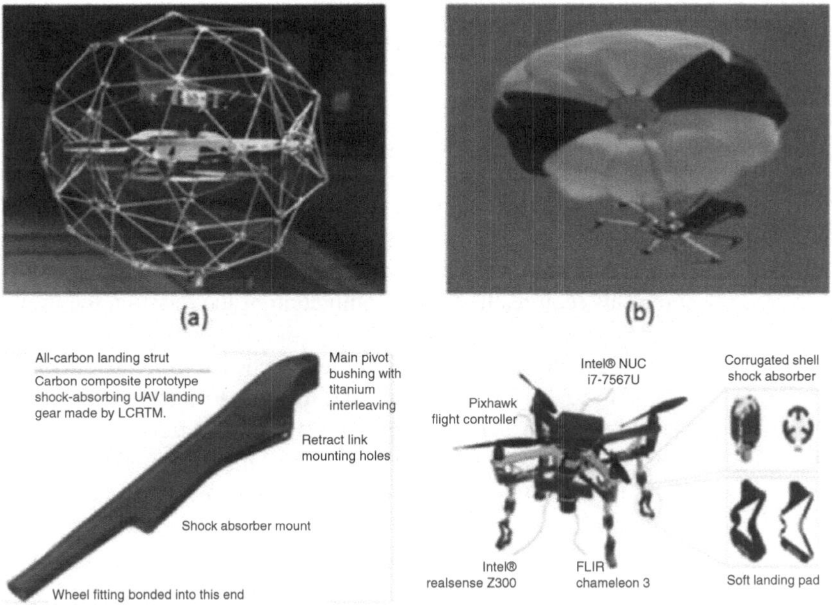

Figure 2.18 Methods available in the market to improve impact performance of landing gear: a) crash-proof multi-copter, b) drone parachute system, c) all-carbon landing structure, d) bio-inspired aerial vehicle design. (Zhang, 2019.)

of a drone required?", "How can precision landing of a UAV be achieved?", and "What are different methods to achieve precision landing?"

Smoothly and precisely landing a UAV in a pre-designated location or a location indicated by the ground pilot or on a location having a unique visual or electromagnetic signature is an important operation performed in the final phase of a UAV. Safe and smooth landing without any crash is important because it allows the UAV to be reusable for successive missions. Besides its own reuse, precisely landing at a desired location is important and highly desirable from different application perspectives. The reliability of a successful UAV mission depends upon autonomous precise landing of the UAV, eliminating any catastrophic consequence endangering the life and resources. Listed below are some of the important applications which require precision landing of a UAV.

a. Package delivery: delivery from stores or critical consignment to precise locations, such as delivery of an Amazon package which belong to the person who has ordered the package.

b. Precision agriculture: sowing seed according to precise row and column of a large field, or precisely spraying pesticide on agricultural land for optimal use of pesticide.
c. Surveillance: precise surveillance of a high-value place or target continuously or intermittently requires pointing the sensors mounted in the belly of the drone at the target, and also that the field of view (FOV) of the sensor precisely locates the target despite turbulence of the UAV or movement of the object under consideration.
d. Drone in a box: often a drone is launched from a box or a canisterized enclosure. This requires that upon completion of the mission, the drone is returned back to the box.
e. Changing of launch station: often a drone is launched from the deck of a ship in order to surveillance the surrounding areas to detect the presence of any hazardous objects floating on the sea surface. After the launching of a drone, the ship is in motion. Therefore, the location of the launch from the ship deck has changed. The drone has to return back to the precise location of the ship deck after performing the surveillance. In order to land back on the ship deck from which it was launched, it must identify the relative position of the ship and the location where it has to land.
f. Indoor landing and navigation: often, a drone operates in an indoor environment where the GNSS signal is denied or not available, such as inside a tunnel or in an indoor stadium or inside a mine. Therefore, there is little room for the drone to manoeuvre and land. It is required that in such scenarios, the UAV be able to take off precisely and land from locations pre-designated for inspection, and there is a little room for any error.

This subsection tries to answer the question "How is precision landing achieved?"

Often, UAV missions are launched, guided, and navigated until they have a safe landing using GNSS. Therefore, these GNSS/GPS-assisted navigations or landings are considered precise landings. However, the precision of locations of any GPS system can have an error of 10 feet in horizontal and vertical positioning. Therefore, the error in landing location of UAVs assisted by GNSS positioning is in a volume of a 10-feet radius. This spherical error in space results in a circular error of probable landing in 2D plane, which is quite high and often catastrophic for landing of a UAV. A probable landing radius of 10 feet is not precise for a UAV landing. What are considered precision landings of a UAV are detected based on the nature of UAV missions and are often less than half a meter circular of radius. Therefore, for a precision landing, the circular or volume error should be less than 10 inches. So a precision landing of a UAV can be defined as a landing where the circular error of landing position does not exceed a circle of radius 10 inches.

2.9 METHODS FOR PRECISION LANDING

There are several methods which are being researched and realized for achieving precision landing of UAVs. Generally, the methods are a combination of hardware, software, and sensors realized through specific algorithms. In this section, two prominent methods used for precise landing of UAVs are discussed along with their advantage and disadvantages.

2.9.1 Method-I

This method is a camera empowered by CV algorithms to identify the image of the location on the ground where the UAV needs to land. The camera quickly identifies and captures the image of the location from a vertical height of about 20/30 ft (10 metres) and compares it with the image pre-fed into the memory of the UAV. A comparison of the images is performed using Scale Invariant Feature Transform (SIFT) points computed from the images. The UAV starts manoeuvring towards the image with precise matching of the SIFT features of the image on the ground, thereby achieving a precise landing.

In this method, the instruments required are

a) Using an optical camera to capture the image on the ground and an algorithm to compute the distance from the camera to the ground image (D)
b) Tracking and capturing the image and computing its SIFT parameters successively.
c) Comparing and matching the SIFT points of the ground image with the SIFT points pre-fed and stored in the memory of the UAV.
d) If SIFT ground image matches with SIFT on-board memory, then manoeuvring towards the ground image.
e) If distance (D < 5 metre), slowing the descent speed through activation of reverse thrust and opening the landing gear until it touches the ground image or location.

The advantage of this method is that it is a low-cost solution, because a camera embedded by CV algorithms for on-board computing of distance from ground image and SIFT point of successive image is available at a cost of approximately ($200 ~ INR 1600). The disadvantage of this method is that it only works within a range where Line of Sight (LoS) vision is available. That is, there must not be any obstruction between the Field of View (FoV) of the camera and the image of the position on the ground.

There should not be any atmospheric disturbance such as fog or dust obstructing the fidelity of the image captured by the camera.

This method does not work in night conditions, and the homing and landing can be achieved from 20–30 ft above ground.

2.9.2 Method-II (Range of altitude as function of the size of the ground image/marker)

In this method, an Infrared (IR) lock for locking to a IR-emitting beacon is employed to achieve a precise landing. An IR camera (~$120), specially fitted with an IR lens and the ground beacon, emits a beacon of IR light, which is powered by a battery or DC power supply.

In the IR-lock method, the ultimate range of the IR lock is 0–60 ft, that is, ~20 metres, which depends on the range of the IR lock solutions. This solution can work from a much higher altitude depending on the range of the IR range finder. The advantages of this method are the following:

The IR lock is a plug-and-play solution with an IR beacon and an IR range finder at an approximate cost of about $500 ~INR 40,000.

A precision landing depends upon the range of the IR beacon. The typical distance of the IR range finder or beacon is ~60" to ~100". This method can perform precision landing in both day and night conditions or in rough or clear weather in windy and foggy atmospheric conditions.

QUESTIONS AND ANSWERS

1. Describe the basic structure of a UAV with its various components and list the functions of each of the components of an UAV.

Answer: The basic structure of a UAV is mechatronic structure with suitable avionic integrated into it. A propeller, which is the mechanism for take-off, flying, and landing of UAVs, is discussed. Also, a UAV has a structure for holding and controlling sensors for navigation and payload sensors for sensing the environment. Each UAV is designed for a specific mission. The mission objectives guide the crafting and designing of the UAV with a specific structure, avionic, propulsion system, and payload sensors out of the many available choices.

The aerodynamic structure of the UAV is embedded with basic functional components such as its propeller, landing gear, avionic, flight control structures, radar, communication and data link to the GCS, gimbal to mount the payload sensors.

Google Answer: An Unmanned Aerial Vehicle (UAV), also known as a drone, is an aircraft that operates without a human pilot on board. The basic structure of a UAV typically includes the following components:

- Fuselage: The fuselage is the main body of the UAV, which houses all the other components. It is usually made of lightweight materials like carbon fibre, aluminium, or plastic.
- Wings: The wings provide lift to the UAV and enable it to fly. They are usually made of lightweight materials and are designed to be highly aerodynamic.
- Propulsion system: The propulsion system includes the motor, propellers, and batteries. The motor drives the propellers, which provide the thrust to the UAV. The batteries power the motor and other electronics on the UAV.
- Navigation system: The navigation system includes sensors like GPS, altimeters, and gyroscopes. These sensors help the UAV maintain its position and altitude, as well as navigate to specific waypoints or follow a predetermined flight path.
- Communication system: The communication system includes the radio transmitter and receiver, which allow the UAV to communicate with the ground station and receive commands from the operator.
- Payload: The payload is the equipment or cargo that the UAV is carrying. This can include cameras, sensors, weapons, or other types of equipment.

The functions of each component of a UAV are as follows:

- Fuselage: The fuselage provides a housing for all the other components and helps maintain the structural integrity of the UAV.
- Wings: The wings provide lift to the UAV and enable it to fly. They are designed to be highly aerodynamic to minimize drag and maximize efficiency.
- Propulsion system: The propulsion system provides the thrust that allows the UAV to move through the air. The motor, propellers, and batteries work together to power the UAV.
- Navigation system: The navigation system helps the UAV maintain its position and altitude, as well as navigate to specific waypoints or follow a predetermined flight path. The sensors in the navigation system provide data to the flight controller, which adjusts the UAV's flight path accordingly.
- Communication system: The communication system allows the UAV to communicate with the ground station and receive commands from the operator. This enables the operator to control the UAV and receive data from its sensors.

- Payload: The payload is the equipment or cargo that the UAV is carrying. This can include cameras, sensors, weapons, or other types of equipment. The payload enables the UAV to perform specific missions, such as surveillance or reconnaissance.

2. What are different components of the GCS of an UAV? Who manages a GCS?

Answer: The different components of the GCS of a UAV are:

- Geometry and structural design of the GCS.
- Hardware components.
- Software components.
- Data link.
  GCSs are of different types, mainly depending on the type of UAV and its mission plan. Some variants are: small handheld remote controller, simple laptop computer, mobile vehicle housing multifunctional display, and command and communication stations. Antennae for transmitters and receivers (Tx-Rx) are mounted above the rooftop of the mobile enclosure for a constant data and command link with the UAV.

Google Answer: The Ground Control Station (GCS) of an unmanned aerial vehicle (UAV) is typically managed by a team of trained professionals. The specific roles and responsibilities of the team members may vary depending on the size and complexity of the operation, but generally the team may include:

- Pilot-in-command (PIC): The PIC is responsible for the safe and legal operation of the UAV. They are the primary operator of the GCS and are responsible for executing the flight plan and monitoring the aircraft's performance.
- Payload operator: The payload operator is responsible for managing the UAV's payload, such as cameras, sensors, or other specialized equipment. They ensure that the payload is functioning correctly and that data is being collected and transmitted back to the GCS.
- Mission coordinator: The mission coordinator is responsible for planning and executing the overall mission, including flight planning, risk assessment, and communication with other members of the team.
- Communications specialist: The communications specialist is responsible for maintaining communication between the UAV and the GCS. They ensure that the data link is functioning correctly and troubleshoot any issues that may arise.
- Maintenance technician: The maintenance technician is responsible for ensuring that the UAV is in good working order. They perform preflight checks and routine maintenance tasks to keep the aircraft in optimal condition.
- Overall, the GCS team works together to ensure the safe and efficient operation of the UAV and its mission.

3. What is a ground pilot? What are the functions of a ground pilot in a UAV mission? List the hardware components.

Answer: The ground pilot or the GCS is a subsystem of the UAV. GCSs are of different types, which are mainly dependent on the type of UAV and their mission plan. Some variants are: small handheld remote controller, simple laptop computer, mobile vehicle housing multifunction display, and command and communication stations. Antennae for transmitters and receivers (Tx-Rx) are mounted above the rooftop of the mobile enclosure for a constant data and command link with the UAV.

The GCS consists of the following hardware components:

- Display screens: they are used to display the payload status, data, and communication status in addition to the altitude and position coordinates of the UAV on a digital map background.
- Control sticks (avionic control and payload control): they are used to deflect the control surfaces and engine throttle of the UAV. Generally, the payload control stick is used to control the operations of payloads like camera or ranging sensors. The sensor control is to orient the focus of the sensor to the particular location of an object. This operation is known as geopointing and geolocation of the payload sensor.
- Keypads: they are used to sending the input mission data for the UAV from the GCS. Data link has two components known as uplink and downlink. The uplink is often in two non-interfering and non-overlapping signal channels. The uplink channel is used to send command and information from the GCS to the UAV.
- The downlink is used to send information from the UAV to the GCS (e.g. graphic image, sensor data such as image frames, video or laser point cloud of the sensed space). The transmission channel consists of data terminal equipment and a wireless trans receiver. A modem, controllers, and cryptographic equipment are the components of the data terminal equipment. The uplink and downlink channels are dynamically encrypted so as to avoid spoofing and hijacking of the UAV by unauthorized operators.

4. What are different hardware and software components of a GCS?

Answer:

- Hardware components: Display screens: they are used to display the payload status, data, and communication status in addition to the altitude and position coordinate of the UAV on a digital map background.
- Control sticks (avionic control and payload control): they are used to deflect the control surfaces and engine throttle of the UAV. Generally, the payload control stick is used to control the operations of payloads like camera or ranging sensors. The control is to orient the focus of the

sensor to a particular location of an object. This operation is known as geopointing and geolocation of the payload sensor.
- Keypads: they are used to send the input mission data for the UAV from the GCS. Data link has two components, known as uplink and downlink. The uplink is often in two non-interfering and non-overlapping signal channels. The uplink channel is used to send commands and information from the GCS to the UAV.
- The downlink is used to send the information from the UAV to the GCS (e.g. graphic image, sensor data such as image frames, video or laser point cloud of the sensed space). The transmission channel consists of data terminal equipment and the wireless channel. A modem, controllers, and cryptographic equipment are components of the data terminal equipment. The uplink and downlink channels are dynamically encrypted so as to avoid spoofing and hijacking of the UAV by unauthorized operators. Software components: The stack of software of the GCS consists of the OS, the mission planner, the mission controller, and post-mission data analyser. Software like mission planner loads waypoints to the autopilot board to configure missions for the UAV. Alternate mission plans based on the UAV log data are also configured and uploaded by the mission planner software.

5. What are different GNSS systems that prevail and how do satellite-based GNSS systems help in navigating and locating the UAV's position?

Answer: The global navigation satellite system (GNSS) is based on a position calculation method on the Earth's surface from pseudo-distance data obtained from a minimum of three known satellite positions, and the presence of a fourth satellite also aids in calculation of the altitude of the UAV.

6. What are different components of a GNSS system?

Answer:

- On-ground auxiliary system and constellation of satellites
- GNSS receiver
- Augmentation system
  - Air-based augmentation system
  - Ground-based augmentation system
  - Satellite-based augmentation system

7. What is an IMU? What are the outputs of an IMU? How can an IMU help in navigating a UAV in the absence of a GNSS signal?

Answer: IMU stands for inertial measurement system, which is designed for a wide range of platforms including UAVs (unmanned aerial vehicles),

UGVs (unmanned ground vehicles), and USVs (unmanned surface vessels). An IMU provides access to all major GNSS constellations and frequencies (GPS L1/L2, GLO L1/L2, GAL E1/E5, BDS B1/B2, and QZSS L1/L2) to initiate the starting position of the UAV. It provides highly accurate navigation in GNSS-denied environments, by utilizing Inertial Labs' new enhanced Tunnel Guide algorithm. The INS-BU delivers a comprehensive GNSS-denied solution.

There is an on-board computer present that processes speed and altitude along with information provided by the sensors in order to obtain the current location from a predefined starting point.

The outputs of IMUs are horizontal and vertical positions, pitch and amp, roll, velocity, accelerations, angular rates, barometric pressure data, and PPS. Also, it sends the data as an NMEA message with position, heading, pitch and amp, and roll to the UAV and the GCS.

8. What are the different elements of an IMU?

Answer:

Elements of an IMU:

- Air data, altitude, and heading reference
    - Barometer
    - Static pitot system
    - Thermometer
- Inertial Measurement Unit
    - Sensor and Supporting electronics
- Gyroscope
- Accelerometer
- Magnetometer

9. What are different output formats of GPS and GNSS systems? What is the NMEA format? Explain various NMEA sentences.

Answer: NMEA sentences are not explained in this chapter:

Answer: GPS and GNSS systems have different output formats, depending on the manufacturer and the specific device. However, the most commonly used output format is the NMEA format. NMEA stands for National Marine Electronics Association, which developed a standard protocol for communication between marine electronic devices.

The NMEA format is used for data transmission between GPS or GNSS receivers and other devices such as chart plotters, computers, and autopilots. The NMEA data is transmitted over a serial port, and it consists of ASCII characters.

There are several types of NMEA sentences, each with a specific function. Some of the most common NMEA sentences include:

GGA (Global Positioning System Fix Data): This sentence contains information about the current position, the number of satellites in view, the accuracy of the position, and the time.

RMC (Recommended Minimum Navigation Information): This sentence provides information about the position, speed, and course of the device.

GSA (GNSS DOP and Active Satellites): This sentence contains information about the GNSS system in use, the number of satellites being tracked, and the dilution of precision (DOP).

GSV (GNSS Satellites in View): This sentence provides information about the satellites in view, their elevation and azimuth, and the signal strength.

VTG (Course Over Ground and Ground Speed): This sentence provides information about the speed and direction of the device over the ground.

ZDA (Time and Date): This sentence provides information about the current date and time in UTC.

In summary, the NMEA format is a standard protocol used for communication between GPS or GNSS devices and other devices. It consists of ASCII characters and includes several types of sentences, each providing specific information about the device's position, speed, and course.

10. Illustrate the networked view of an UAV with its possible communication links (Figure 2.19).

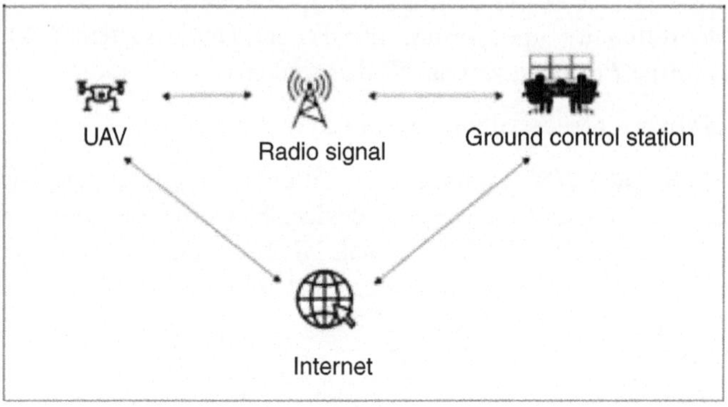

Figure 2.19 Networked view of an UAV with its possible communication links.

11. What are some different communication technologies used to communicate between a GCS and a UAV?

Answer: Wi-Fi, Bluetooth, Zigbee, WiMax

12. List all the short-range and long-range communication technologies used in UAV communication.

Answer:

Short range: Wi-Fi, Bluetooth, Zigbee, WiMax

Long range: long-term evolution (LTE), cellular mobile communication, satellite communication (SATCOM)

13. How can communication from satellites help a UAV mission? Illustrate with a diagram the communication components between a UAV and a GCS (Figure 2.20).

Answer: A UAV to GNSS or a satellite, which is a unidirectional link from satellite to a UAV receiving navigational information. This link from a UAV to a satellite is used for sending real-time navigation data with the help of GNSS or GPS.

14. What are different hybrid communication technologies used in a UAV mission?

Answer: Hybrid communication systems used for UAVs:

- Cellular and wireless hybrid
- Bluetooth and cellular

15. What are different technical challenges faced by UAV communication and what are possible solutions?

Answer: Challenges of UAV communication and possible solutions are as follows:

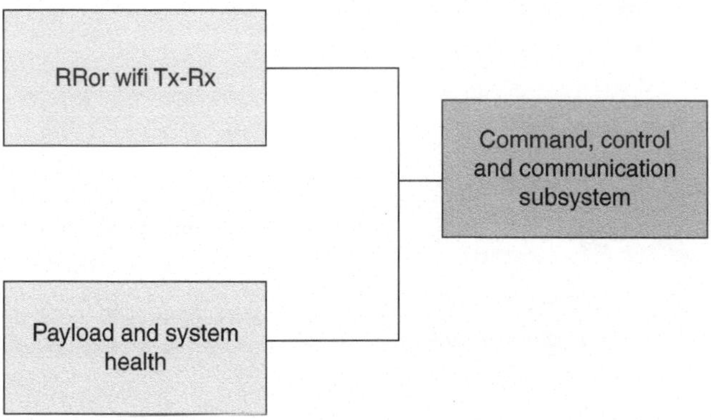

Figure 2.20 Communication components between an UAV and GCS.

- Multipath signal fading in a large class of UAVs used for civilian purposes

  The signal transmitted between the UAV and the GCS is required to be highly reliable and robust because it would pose a danger to other aircraft in flight or humans on the ground. Some of the parameters used to check the performance of the communication system are the following:
  - Availability: it is defined as "the probability that an operational communication transaction can be initiated when needed."
  - Integrity: Refers to "the probability that communication transactions are completed within the communication transaction time with undetected error."
  - Continuity: Refers to "the probability that an operational communication transaction can be completed within the communication transaction time."
  - Latency and communication transaction time: Latency is the maximum time taken for a one-way-message to be delivered. Communication transaction time is the time taken for completion of the operational communication transaction.
  - Inclusion of the human response time

16. List all the different landing gear mechanisms used in different types of UAVs, and in drones.

Answer:

(a)

Simple vertical structure supporting individual rotor arm at the rotor end

(b)

Four vertical structures connected to the centre of the chassis frame

(c)

Vertical structure attached to the centre of the chassis with two vertical structures connected with a horizontal member

(d)(www.unmannedsystemstechnology.com/ /)

Two vertical members connecting one end to the centre of the chassis frame and the other end to horizontal skid-like members

(e)(https://cfm-air.net/projects/drone-multicopter/)

Three vertical members connected to the centre of the chassis frame consisting of horizontal modifications for support at the other end

17. What is a precision landing and how it is achieved in the landing of a UAV?

Answer: Smoothly and precisely landing a UAV in a pre-designated location or a location indicated by the ground pilot or on a location having a unique visual or electro-magnetic signature is an important operation performed in the final phase of a UAV. A safe and smooth landing without any crash is important because it allows the UAV to be reusable for successive missions.

Method used for precision landing:

Camera empowered by CV algorithms to identify the image of the location on the ground where the UAV needs to land. The camera quickly identifies and captures the image of the location from a vertical height of about 20/30 ft (10 metres) and compares it with the image pre-fed into the memory of the UAV.

The comparison of the images is performed using Scale Invariant Feature Transform (SIFT) points computed from the images. The UAV starts manoeuvring towards the image with precise matching of the SIFT features of the image on the ground, thereby achieving a precise landing.

REFERENCES

Barrows, G., Chahl, J., Srinivasan, M. Biologically Inspired Visual Sensing and Flight Control. *Aeronaut. J.* 2003, 107(1069), 159–168. 10.1017/S0001924000011891

Franta, N.R. Swarm Intelligence for Autonomous UAV Control. Naval Post Graduate School, Monterey, California; 2005 June.

Gageik, N. et al. Waypoint Flight Parameter Comparison of an Autonomous UAV. *Int. J. Artificial Intelligence Appl.* 2013 May, 4(3).

Geraci, G., Garcia-Rodriguez, A., Giordano, L.G., Lopez-Perez, D., Bjornson, E. Understanding UAV Cellular Communications: From Existing Networks to Massive MIMO. *IEEE Access*. 2018, 6, 67853–67865.

Hildmann, H., et al. Nature-Inspired Drone Swamming for real time Data Collection under Dynamic Operational Constraints. *MDPI*. 2019 Sept.

Li, B., et al. UAV Communications for 5G and Beyond: Recent Advances and Future Trends. *IEEE J.* 2019, 6(2).

Li, B., Fei, Z., Zhang, Y. UAV Communications for 5G and Beyond: Recent Advances and Future Trends. *IEEE Internet Things J.* 2018, 6, 2241–2263.

Liu, Fei et al. "A Real-Time and Multi-Sensor-Based Landing Area Recognition System for UAVs." *Drones* (2022): n. pag.

Mozaffari, M., Saad, W., Bennis, M., Nam, Y.-H., Debbah, M. A Tutorial on UAVs for Wireless Networks: Applications, Challenges, and Open Problems. arXiv 2018, arXiv:1803.00680.

Oubbati, O.S., Atiquzzaman, M., Ahanger, T.A., Ibrahim, A. Softwarization of UAV Networks: A Survey of Applications and Future Trends. *IEEE Access*. 2020, 8, 98073–98125.

Sadraey, M.H. Design of Unmanned Aerial systems, First Edition. John Wiley and Sons Ltd; 2020.

Ullah, Z., Al-Turjman, F., Mostarda, L. Cognition in UAV-Aided 5G and Beyond Communications: A Survey. *IEEE Trans. Cogn. Commun. Netw*. 2020, 6, 872–891.

Zhang, K., Chermprayong, P., Tzoumanikas, D., Li, W., Grimm, M., Smentoch, M., Leutenegger, S., Kovac, M. Bioinspired Design of a Landing System with Soft Shock Absorbers for Autonomous Aerial Robots. *J. Field Robotics*. 2019, 36(1), 230.

Zhang, S., Zhang, H., Di, B., Song, L. Cellular UAV-to-X Communications: Design and Optimization for Multi-UAV Networks. *IEEE Trans. Wirel. Commun*. 2019, 18, 1346–1359.

Zhi, Y., Fu, Z., Sun, X., Yu, J. Security and Privacy Issues of UAV: A Survey. *Mob. Netw. Appl.* 2019, 25, 95–101.

Chapter 3

Designing the fuselage, wing, propulsion, and empennage of a fixed-wing UAV

OVERVIEW

The design process of an aircraft is an iterative process of model-analysis-test both at the subsystem level and the system level, resulting in an aircraft which is optimized for the required performance, stability, and mission (Mohammad, 2020). The iterative design process can be automated with the help of various tools and techniques such as Computer-Aided Design for structural design of the craft, Computational Fluid Dynamics Analysis for moving geometry analysis, structural analysis, crash impact analysis for endurance, impact analysis of landing gear and structure, network and communication simulation for link failure, channel capacity analysis, and so on (Jaeger et al., 2013).

This section discusses some of the important design aspects of the craft, namely, its wing, fuselage, propulsion system, and empennage. Also discussed are different constraints in designing a fixed-wing Unmanned Aerial Vehicle (UAV). The different landing gears used in VTROL UAV are discussed.

3.1 GENERAL LAYOUT OF A FIXED-WING UAV

Analysing the design of existing configurations of a fixed-wing Unmanned Aerial Vehicle (UAV) helps us in generating lots of ideas about the size, and wing characteristics like taper ratio, aspect ratio, and the span of the craft. It also be helpful in understanding the common layout of a UAV and its subsystem

The diagram in Figure 3.1 indicates the common layout of a fixed-wing UAV. The payload used in the UAV is dependent on the mission requirements, and the complexity of the electronic components is again a factor in its application. A battery-powered UAV is shown in the figure with a propulsion system like a jet combustion engine, a navigation system, and other avionic subsystems used in the UAV.

DOI: 10.1201/9781003432685-3

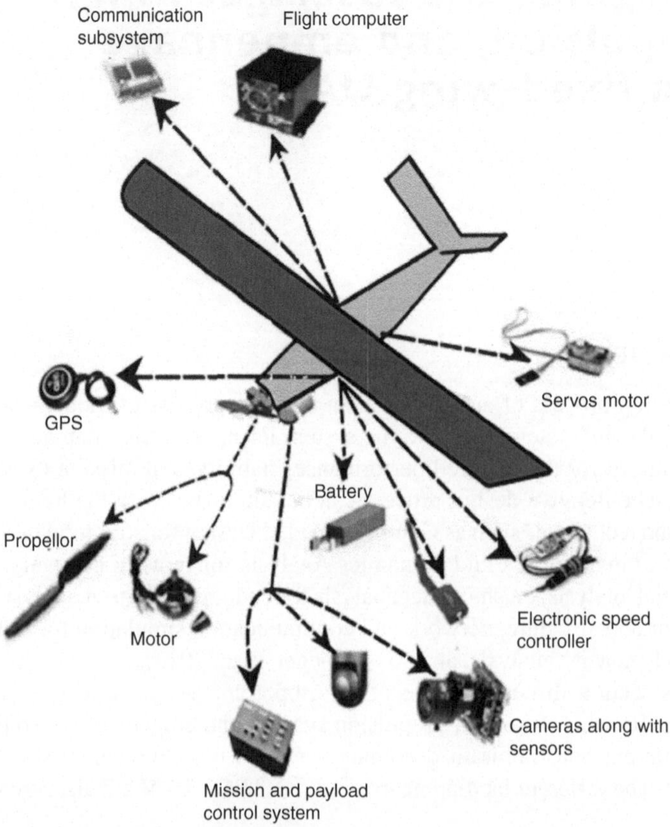

Figure 3.1 General layout of a fixed-wing UAV.

3.2 CONSTRAINT DIAGRAM

The constraint diagram was first established by NASA and was later developed by Roskam. The constraint diagram essentially reduces the solution space and results in an optimized value of T/W and W/S (power loading and wing loading). The conceptual design of the UAV is done by generating the power loading versus the wing loading for a fixed-wing UAV which is jet propelled or propeller based (Ashraf et al., 2018). The UAV is subjected to the constraints that are put forward by the performance requirements of the UAV. Some of the major performance requirements are maximum velocity, absolute vertical ceiling, rate of climb, rate of turn, stall velocity, and take-off and landing roll distances. A sample constraint diagram is depicted in Figure 3.2.

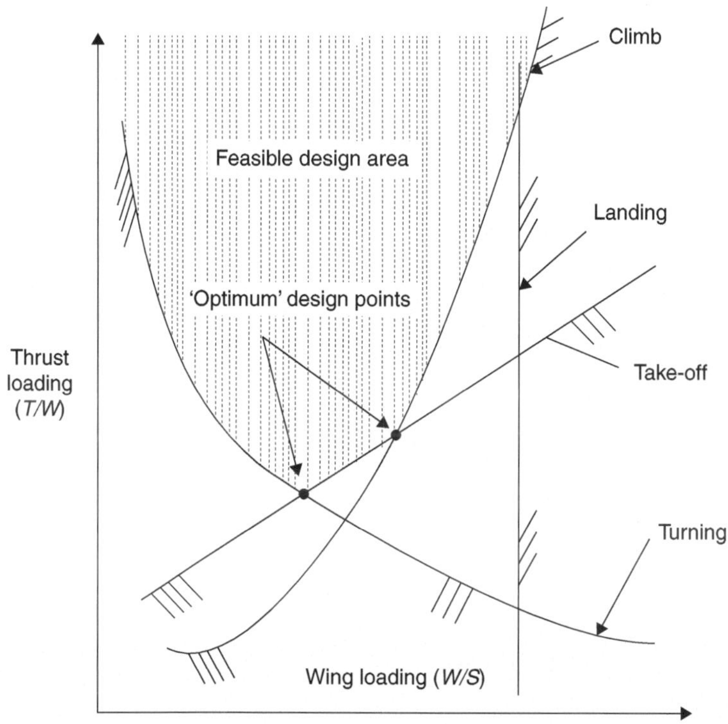

Figure 3.2 Constraint diagram. (www.researchgate.net/figure/Figure-9-Generalised-Constraint-Diagram_fig7_280933114.)

3.3 WING DESIGN

The wing loading obtained from the constraint diagram is used to determine the wing area, with the use of maximum take-off weight. The maximum take-off weight has to be estimated prior to the design of the constraint diagram. The wing configuration of the UAV (Figure 3.3) is mapped to one of the designed structures from the survey of available configurations of UAVs, namely, high wing, mid wing, and low wing (El Adawy et al., 2023).

After producing the constraint diagram, the following steps are used to obtain the wing characteristics:

- Wing area determination
- Calculation of the lift coefficient required by the whole aircraft, wing and the airfoil at cruise conditions
- Selection of airfoil based on the Reynolds number and the coefficient if lift required by the airfoil at cruise conditions

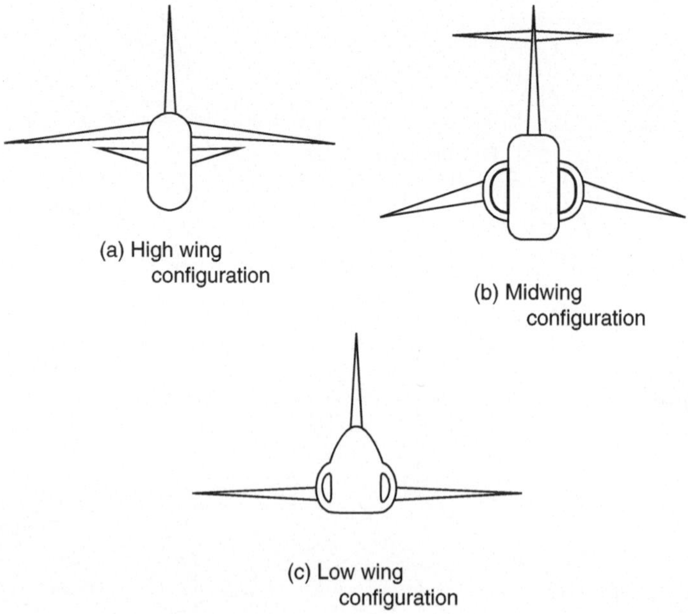

Figure 3.3 Wing configurations.

- Selection of the wing plan from some of the examples are elliptical rectangular tapered and sweptback wing
- Selection of the wing parameters like aspect ratio, taper ratio, sweep, and taper based on the survey data or by analysing the flow patterns using 3D simulations
- Calculation of wing geometrical characteristics like mean aerodynamic chord (MAC), wing span, root chord, tip chord using Computational Fluid Dynamics (CFD) codes like Lateral Load Transfer (LLT) and the panel code (Dhekane and Sherje, 2020)
- Selection of the dihedral or anhedral angle from the stability analysis
- Calculation of the Reynolds number at the cruise based on the stability requirement

3.4 SELECTION OF THE PROPULSION SYSTEM

From the constraint diagram, T/W or P/W is obtained and the Maximum Take-Off Weight (MTOW) is assumed or roughly estimated. From the MTOW, the thrust required for a jet engine-based propulsion system or the power required for the propeller-based propulsion system can be calculated. On determining the thrust or power requirements, a suitable power plant to meet the requirement is searched for. In the case of a jet-based propulsion system, the optimization is done such that the thrust-specific fuel consumption is low, weight is low, and the thrust output is high. For a propeller-based system, maximum power with a low-weight system is preferred. This is the reason for selecting a brushless motor instead of a brushed motor (Cha et al., 2021).

There are propulsion systems available that optimize software like Moto Calc or eCalc to obtain performance based on the specifications of the motor, Electronic Speed Controller (ESC) battery, and the propeller selected.

Battery selection can be done depending on the cost, energy density, and the MTOW of the UAV. Various energy densities for different powerhouses are displayed in the Ragone plot that is shown in Figure 3.4. From the figure

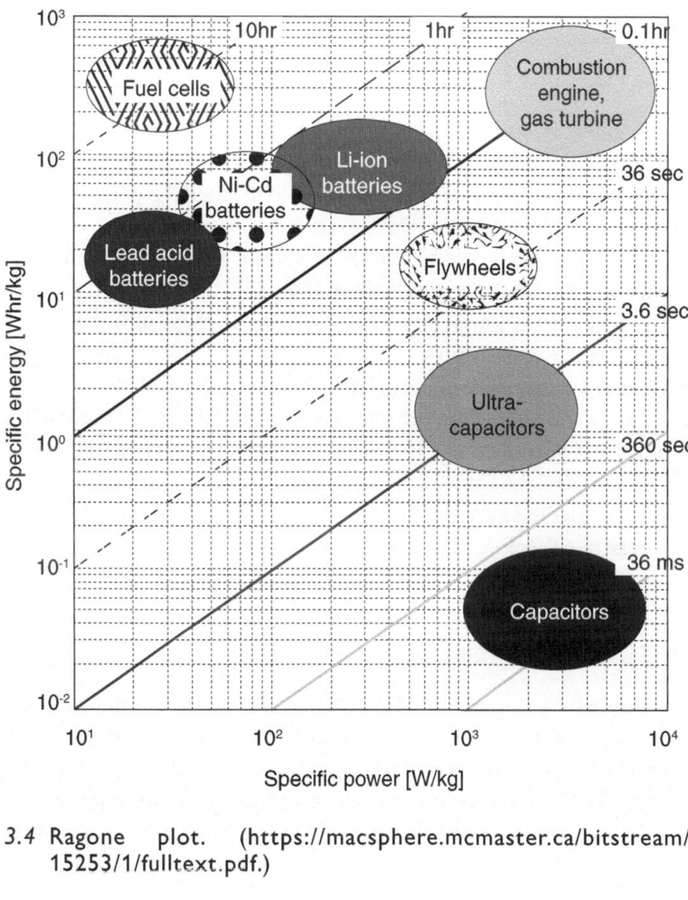

Figure 3.4 Ragone plot. (https://macsphere.mcmaster.ca/bitstream/11375/15253/1/fulltext.pdf.)

we can observe that fuel cell has highest energy density in comparison to Li-Po batteries, which are better than Ni-Cad or Ni-MH batteries. Li-Po batteries also have a better power density, that is, readiness to discharge fast as compared to fuel cells or Ni-Cad or Ni-MH batteries. The figure also indicates the endurance for a constant discharge use in robots and UAVs, and it can be inferred that Li-ion batteries have nearly an hour of endurance, while fuel cells have around 10 hours.

The battery ratings provided on Li-Po batteries indicate the constant and burst current discharge the battery is capable of. The ratings are provided as 25C–50C and 15C–60C. The first number indicates the constant discharge capacity, while the latter indicates the burst current capacity. The current discharge for a 2,000 mAh battery with a rating of 25C–50C can be calculated as follows:

$$\text{Constant current } I_C = (2000/1000)*25 = 50A$$

$$\text{Burst Current } I_B = (2000/1000)*50 = 100A$$

Based on the constant current discharge, the battery endurance can be concluded to be (2000*3600/1000)/50 =144s = 2.4 minutes.

If the constant current drawn by the motor is 10A, then the endurance improves to 720 seconds, or 12 minutes.

The propeller for the motor is chosen based on the recommended propeller size provided by the motor manufacturer. The ESC for the motor is chosen in such a way that the ESC has at least a 10A to 15A higher current rating than the motor maximum current to protect the ESC. It is recommended to choose an ESC with battery eliminator circuit (BEC) that uses an ESC to power the control surface servos through the receiver. Without a BEC, an external BEC can be used in parallel with the ESC to power the receiver, servo, and other avionics (Wagter et al., 2020).

The avionics selected are modelled along with their weights, which shall later be used to generate fuselage based on Centre of Gravity (CG) location and wing geometry.

3.5 FUSELAGE DESIGN

The fuselage design is performed based on the payloads that are used and the avionics used in the UAV. The payloads and avionics are arranged in the fuselage to get the CG at the required position, and the arrangement is wrapped with surfaces. If the UAV is very small, a monocoque fuselage can be used. If not, a boom is used to hold the empennage section in position. The boom is a structure that helps in reducing the drag of the fuselage but with compromise in the structural strength. Ease of fabrication and aerodynamic performance with less drag need to be considered when designing a fuselage. The fuselage is analysed in CFD tools to obtain characteristics like that moment coefficient that would be helpful in tail design. The structural performance can also be analysed using Finite Element Method (FEM) computations (Shahriar et al., 2014).

The three major configurations of the fuselage are the following:

5.1 **Box type:** The box type fuselage extends from the firewall where the motor is mounted to the aft tail. It tapers towards the empennage section to push the aircraft CG forward. The box structure can be either made of strong skin called the monocoque or of truss members covered by a skin called truss (Nicolosi et al., 2021).

5.2 **Single Boom:** In this type of fuselage, the payload is wrapped in an aerodynamically smooth container. The tail of the UAV is attached to the main fuselage through a single structure called a boom. This fuselage type has lower drag as compared to the box structure, but the strength is also less.

5.3 **Twin Boom:** Twin boom fuselage connects the empennage section to the main fuselage through two booms. This helps in obtaining less as well as better strength capabilities (Vehicle, 2011).

3.6 EMPENNAGE DESIGN

The design of the empennage section is based on the stability aspects of the UAV. The aim is to design an empennage section that stabilizes the UAV in longitudinal as well as lateral directions while generating low drag at cruise conditions (Figure 3.5). The design process begins with

- Selection of the tail volume coefficient for both horizontal and vertical tails
- Estimation of tail arm length for both the tail surfaces
- Determination of tail surface area for the horizontal and vertical tail
- Calculation of the tail geometries like root and tip chord, span and taper ratio
- Computation of lift required by horizontal tail surface to achieve longitudinal static stability
- Selection and analysis of airfoil for both the tail surfaces
- Analysis of stability to ensure longitudinal and lateral stability (Zuliardiansyah and Rasyiddin, 2022).

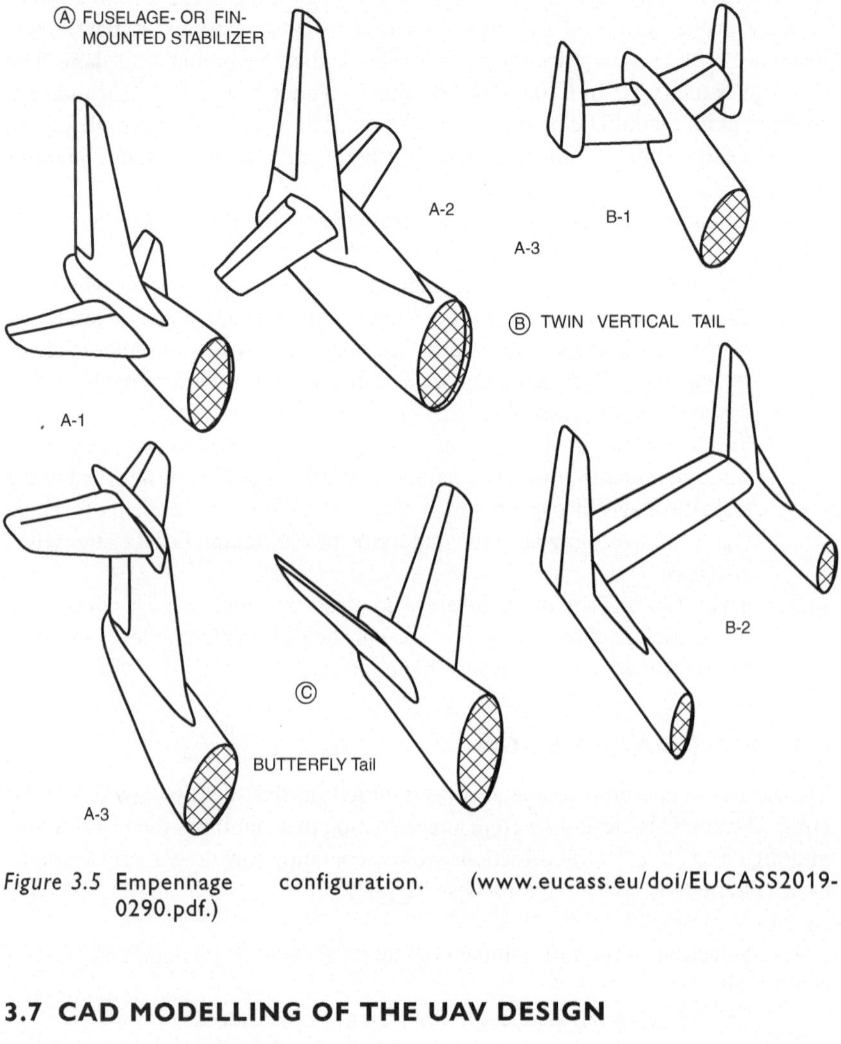

Figure 3.5 Empennage configuration. (www.eucass.eu/doi/EUCASS2019-0290.pdf.)

3.7 CAD MODELLING OF THE UAV DESIGN

The Computer-Aided Design (CAD) model of the aircraft is generated mainly to extract the CG of the UAV based on the compact arrangement of payloads and subsystems. The CAD model will also aid in simulating the positioning of the payloads and avionics which can be used to create a compact fuselage.

Subsystems like the wing, empennage, and propulsion systems are placed, and the arrangement objects inside the fuselage are varied to obtain the CG of the UAV at the desired location. The CAD model created during the preliminary phase of design need not be detailed as required during the fabrication phase. The model should contain the details of geometry and mass of the UAV components, which will provide information of CG, and the mass moment of inertia of the whole UAV will be useful for the stability analysis of the UAV (Vasić et al., 2018).

3.8 CFD ANALYSIS OF UAV DESIGN

Once the design of the entire UAV is completed, it is essential to conduct aerodynamic performance testing using numerical techniques. This testing allows for the extraction of values such as lift coefficient, drag coefficient, and moment coefficient about the CG of the aircraft. Numerical analysis enables a detailed study of the flow around the aircraft and its subsystems, leading to improvements in aerodynamic performance. The identification and avoidance of stagnation points based on the complete UAV numerical analysis results contribute to enhanced airflow around the aircraft. Additionally, studying and improving the effects of wing downwash on downstream geometries can be achieved.

The numerical analysis of aerodynamic performance plays a crucial role in evaluating and refining the design of the UAV. By employing CFD techniques, engineers can simulate the airflow around the aircraft and obtain valuable data on various aerodynamic parameters. The lift coefficient, which quantifies the lift generated by the wings, and the drag coefficient, which represents the resistance encountered by the aircraft in the airflow, are fundamental factors in assessing the UAV's performance.

The moment coefficient about the CG of the aircraft is another key parameter obtained from numerical analysis. This coefficient provides insight into the aircraft's stability and control characteristics. It helps in determining the aircraft's response to control inputs and its ability to maintain a desired flight path.

3.8.1 Performance analysis

Performance analysis of an aircraft is a crucial step that is conducted once the design of the aircraft is complete. It plays a vital role in validating the design parameters obtained from the constraint diagram, ensuring that the UAV meets the desired performance requirements. The performance analysis also helps to obtain important performance characteristics of the aircraft, such as range, endurance, and maximum velocity at a particular altitude (Varsha and Somashekar, 2018).

The performance analysis serves as a means to verify the aircraft's capabilities and to assess its ability to fulfil its intended purpose. It involves a detailed evaluation of various factors that impact the aircraft's performance, including aerodynamics, propulsion system, weight and balance, and operational conditions. By analysing these aspects, designers can determine the aircraft's performance limits and make informed decisions regarding its design and operation.

One of the key objectives of performance analysis is to determine the range of the UAV. Range refers to the maximum distance that the aircraft can travel without the need for refuelling or recharging. By considering factors such as fuel capacity, propulsion efficiency, and aerodynamic characteristics, designers can calculate the range of the aircraft. This information is crucial for mission planning and operational feasibility (Varsha and Somashekar, 2018).

3.8.2 Stability analysis

Stability analysis of the aircraft is performed in two stages. The first stage involves confirmation of static longitudinal and lateral stability; the second stage provides the dynamic stability of the aircraft in both modes

Static stability ensures that the aircraft is in equilibrium during cruise conditions. The wing generates a moment along with the generation of lift force. This moment has to be countered by the horizontal tail by the generation of lift force. This moment has to be countered by the horizontal tail by generating lift at certain distance called tail arm length (Kim et al., 2021).

Longitudinal static stability is achieved by the design of horizontal tail. Roll stability is achieved through vertical tail design. The stability of the UAV design depends on the CG location of the UAV, the aerodynamic centre of the wing, wing and tail airfoil characteristics, and tail arm length. A plot of moment coefficient of the UAV and its subsystems like wing fuselage and tail versus the angle of attack is generated, and the slope of the aircraft curve is observed to analyse the static stability of the UAV

Dynamic stability deals with how fast or how the equilibrium is achieved after the equilibrium state is disturbed through perturbations. A biquadratic equation with coefficient matrix containing UAV moment derivatives is solved to extract Eigen values, and vectors provide the mode shapes of the aircraft movement. The aim is to design the empennage section in such a way that the vibrations are overdamped, which will ensure that the aircraft returns to the equilibrium within less time.

QUESTIONS AND ANSWERS

1. What is a VTROL UAV? How does a VTROL differ from a generic UAV?

Answer: VTROL stands for Vertical Take-Off and Landing UAV. A VTROL differs from a generic UAV in that it is capable of taking off and landing vertically, whereas a generic UAV typically requires a runway or catapult launch.

2. Give a component layout of a fixed-wing UAV and describe the functions of the components.

Answer: Designing the fuselage, wing, propulsion, and empennage of a fixed-wing UAV provides a general layout of a fixed-wing UAV, shown in Figure 3.1. The components of a fixed-wing UAV include:

1. Fuselage: The main body of the aircraft that houses the payload and other essential components.
2. Wing: The primary lifting surface that generates lift to keep the aircraft airborne.
3. Propulsion System: The system that provides thrust to propel the aircraft forward.
4. Empennage: The tail section of the aircraft that includes the horizontal stabilizer, vertical stabilizer, and control surfaces such as elevators and rudder.

The functions of these components are as follows:

1. Fuselage: Provides structural support for the aircraft and houses the payload and other essential components such as avionics, fuel tanks, and landing gear.
2. Wing: Generates lift to keep the aircraft airborne and provides stability during flight.
3. Propulsion System: Provides thrust to propel the aircraft forward and maintain its speed during flight.
4. Empennage: Provides stability and control during flight by controlling pitch (elevators), yaw (rudder), and roll (ailerons).

3. What are different types of wing design patterns used in the design of UAVs?

Answer: There are several types of wing design patterns used in the design of UAVs. These include:

- Rectangular Wing: A simple and easy-to-manufacture wing design that provides good lift and stability.
- Tapered Wing: A wing design that gradually decreases in width from the root to the tip, providing better performance at high speeds.
- Swept Wing: A wing design where the leading edge is angled backward, reducing drag and increasing speed.
- Delta Wing: A triangular-shaped wing that provides high manoeuvrability and stability at high speeds.

4. What are the essential design parameters of a wing of a UAV?

Answer: The essential design parameters of a wing for a UAV include:

- Aspect Ratio: The ratio of the wingspan to its mean chord length. Higher aspect ratios provide better lift-to-drag ratios but can be less stable.
- Airfoil Shape: The shape of the cross section of the wing. Different airfoil shapes provide different lift and drag characteristics.

- Wing Loading: The weight of the aircraft divided by its wing area. Higher wing loading requires more lift to keep the aircraft airborne.
- Sweep Angle: The angle between the leading edge of the wing and a perpendicular line drawn from its root chord. Swept wings reduce drag at high speeds but can be less stable.
- Dihedral Angle: The upward angle between each wing tip and a horizontal plane. Dihedral wings provide lateral stability during flight.

5. What are different types of fuselage used? What is a constraint diagram? What is the solution obtained from the constraint diagram of a fixed-wing UAV?

Answer: The constraint diagram is a tool used in the conceptual design of a UAV to generate power loading versus wing loading for fixed-wing UAVs that are jet-propelled or propeller-based. The constraint diagram helps to reduce the solution space and optimize the values of power loading (T/W) and wing loading (W/S) based on performance requirements such as maximum velocity, absolute vertical ceiling, rate of climb, rate of turn, stall velocity, and take-off and landing roll distances. The solution obtained from the constraint diagram is an optimized value for T/W and W/S that meets the performance requirements while minimizing weight and maximizing efficiency.

6. How are CAD models helpful in designing a UAV? Which parts and components of a UAV can be modelled using CAD? List some of the prevailing CAD modelling software used for designing the avionic components of an UAV.

Answer: CAD (Computer-Aided Design) models are helpful in designing a UAV because they allow for the creation of detailed 3D models that can be used to visualize and test different design concepts. CAD models can also be used to simulate the performance of different components and systems, allowing for optimization before physical prototypes are built.

Parts and components of a UAV that can be modelled using CAD include:

- Fuselage
- Wings
- Propulsion system
- Landing gear
- Avionics components

Some of the prevailing CAD modelling software used for designing avionics components of a UAV include:

- SolidWorks
- CATIA

- AutoCAD
- Siemens NX
- PTC Creo

7. What is CFD? How does CFD analysis help in the verification of a UAV model?

Answer: CFD stands for Computational Fluid Dynamics, which is a numerical technique used to simulate the behaviour of fluids (liquids or gases) in motion. In the context of UAV design, CFD analysis can be used to simulate the airflow around the aircraft and predict its aerodynamic performance.

CFD analysis helps in the verification of UAV models by providing information on important performance parameters such as lift coefficient, drag coefficient, and moment coefficient about the CG of the aircraft. This information can be used to optimize the design of the UAV and ensure that it meets performance requirements such as maximum velocity, rate of climb, and range. CFD analysis can also help identify potential issues with the design, such as areas of high drag or turbulence, which can be addressed before physical prototypes are built. Overall, CFD analysis is a valuable tool for verifying and optimizing UAV designs before they are manufactured and tested in real-world conditions.

8. What are the different performance parameters of a UAV structure?

Answer: The different performance parameters of a UAV structure include:

- Maximum velocity: The highest speed that the UAV can achieve.
- Absolute vertical ceiling: The maximum altitude that the UAV can reach.
- Rate of climb: The rate at which the UAV can ascend vertically.
- Rate of turn: The rate at which the UAV can change direction in flight.
- Stall velocity: The minimum speed at which the UAV can maintain level flight.
- Take-off and landing roll distances: The distance required for the UAV to take off or land.

REFERENCES

Cha, J., Yun, J., & Hwang, H. (2021). Initial sizing of a roadable personal air vehicle using design of experiments for various engine types. *Aircraft Engineering and Aerospace Technology*. ahead-of-print. 10.1108/AEAT-04-2020-0073.

Dhekane, T., Sherje, N. (2020). CFD Simulation of Different Taper Ratio Wings, Performing Trade-off Assessment and Development of A New Methodology to Plot Lift Distribution Curve and 3D Local Coefficient of Lift Distribution Graph. *International Journal of Engineering and Technical Research*, 9. 10.17577/IJERTV9IS040471.

El Adawy, M., Abdelhalim, E.H., Mahmoud, M., Abo zeid, M.A., Mohamed, I.H., Othman, M.M., ElGamal, G.S., ElShabasy, Y.H. (2023). Design and Fabrication of a

Fixed-wing Unmanned Aerial Vehicle (UAV). *Ain Shams Engineering Journal*, 14(9), 102094, ISSN 2090-4479, https://doi.org/10.1016/j.asej.2022.102094.

Jaeger, L., Gogu, C., Segonds, S., Bes, C. (2013). Aircraft Multidisciplinary Design Optimization Under Both Model and Design Variables Uncertainty. *Journal of Aircraft*, 50, 528–538. 10.2514/1.C031914.

Kamal, A.M., Ramirez-Serrano, A. (2018). Design Methodology For Hybrid (VTOL + Fixed Wing) Unmanned Aerial Vehicles. *Aeronautics and Aerospace Open Access Journal*, 2(3), 165–176.

Kim, D., Lee, Y., Oh, S., Park, Y., Choi, J., Park, D. (2021). Aerodynamic Analysis and Static Stability Analysis of Manned/Unmanned Distributed Propulsion Aircrafts Using Actuator Methods. *Journal of Wind Engineering and Industrial Aerodynamics*, 214, 104648, ISSN 0167-6105, https://doi.org/10.1016/j.jweia.2021.104648.

Mohammad H, Sadraey (2020), Design of Unmanned Aerial systems, First Edition, John Wiley and Sons Ltd.

Nicolosi, F., Corcione, S., Trifari, V., De Marco, A. (2021). Design and Optimization of a Large Turboprop Aircraft. *Aerospace*, 8(5), 132. https://doi.org/10.3390/aerospace 8050132.

Shahriar, A., Rahman, M., Saeed, A., Hasib, M., Salam, A. (2014). UAV Fuselage Design and Structural Analysis.

Varsha, N., Somashekar, V. (2018). Conceptual Design of High Performance Unmanned Aerial Vehicle. *IOP Conference Series: Materials Science and Engineering*, 376, 012056. 10.1088/1757-899X/376/1/012056.

Vasić, Z., Maksimovic, S., Georgijević, D. (2018). Applied Integrated Design in Composite UAV Development. *Applied Composite Materials*, 25. 10.1007/s10443-017-9611-y.

Vehicle, U.A. (2011). Concept Generation and Selection (Doctoral dissertation, Department of Electrical and Computer Engineering, Florida State University).

Wagter, C.D., Remes, B.D., Ruijsink, R., Tienen, F.V., Horst, E.V. (2020). Design and Testing of a Vertical Take-Off and Landing UAV Optimized for Carrying a Hydrogen Fuel Cell with a Pressure Tank. *Unmanned System*, 8, 279–285.

Zuliardiansyah, G., Rasyiddin, A. (2022). Performance Analysis of Empennage Configurations on a Surveillance and Monitoring Mission of a VTOL-Plane UAV Using a Computational Fluid Dynamics Simulation. *Aerospace*, 9, 208. 10.3390/aerospace9040208.

Chapter 4

Application-specific UAVs and their requirements

OVERVIEW

The Unmanned Aerial Vehicle (UAV) is a multipurpose, multi-utility vehicle often employed for different missions. UAVs are employed to accomplish different missions which are dull, dirty, and enduring operations where it requires persistence and where human life is a risk. In this chapter we discuss some of the important but popular applications of UAVs. The specification of the UAV for different missions and how they are modified or retrofitted specific to the mission are discussed. This gives students and researchers an idea of the different configurations and payloads of UAVs for different missions (Shakhatreh et al., 2019).

The UAV is a multi-utility vehicle employed for different missions which are dirty, dangerous, and durable operations, and where human life is in danger. Some of the applications of UAVs and their generic configuration are discussed in the rest of the chapter (Singhal, 2018).

4.1 APPLICATION OF UAVS FOR LIDAR MAPPING AND PHOTOGRAMMETRY

Photogrammetry and Light Detection and Ranging (LiDAR) mapping are two widely used techniques in the field of geospatial data collection and analysis. With the advancements in Unmanned Aerial Vehicle (UAV) technology, these techniques have found new applications and have become more accessible, efficient, and cost-effective. By integrating LiDAR sensors and cameras on UAV platforms, it is possible to obtain high-quality data for creating 3D models and accurate terrain information.

Photogrammetry, as defined by Colomina (2014), is the science of making measurements from photographs to generate 3D models of real-world land masses or objects. It involves capturing overlapping aerial images from different angles and then processing them to extract accurate measurements and create detailed models. Traditionally, photogrammetry was performed using manned aircraft or satellites, which had limitations in terms of cost, flexibility, and

DOI: 10.1201/9781003432685-4

availability. However, with the advent of UAVs, photogrammetry has become more accessible to a wider range of applications (Nex et al., 2022).

UAVs equipped with high-resolution cameras and specialized software can capture aerial images with high precision and accuracy. These images can be processed using photogrammetric software to generate dense point clouds, orthomosaics, and Digital Surface Models (DSMs). The resulting 3D models can be used in various fields such as urban planning, agriculture, environmental monitoring, and infrastructure development. UAV-based photogrammetry offers advantages such as rapid data acquisition, cost-effectiveness, and the ability to access hard-to-reach or dangerous areas.

LiDAR mapping, on the other hand, utilizes laser scanners mounted on UAVs to measure the heights of points on the ground or objects below the aircraft. LiDAR systems emit laser pulses and measure the time it takes for the reflected pulse to return to the sensor, enabling the generation of highly accurate 3D point clouds. These point clouds can then be used to create Digital Terrain Models (DTMs), Digital Elevation Models (DEMs), and other geospatial products. UAV-based LiDAR mapping offers several advantages over traditional ground-based or manned aircraft LiDAR systems. UAVs can collect data at lower altitudes, resulting in higher point density and improved vertical accuracy. They also offer increased flexibility and manoeuvrability, allowing for efficient data collection in complex terrains or areas with limited accessibility. Furthermore, the use of UAVs reduces costs and improves safety by eliminating the need for ground-based surveys or manned aircraft operations (Debnath et al., 2023).

The integration of LiDAR sensors and cameras on UAV platforms requires modifications to the UAV system and subsystems. These modifications include the installation of payload mounts, power systems, and data storage and transmission capabilities. Specialized software and algorithms are used to process the acquired data and generate the desired outputs, such as point clouds, orthomosaics, and 3D models. Several studies and research papers have highlighted the successful application of UAVs for photogrammetry and LiDAR mapping. For example, Everaerts (2008) discusses the use of UAV-based LiDAR for forest monitoring and inventory, demonstrating the effectiveness of this technique in capturing detailed information about vegetation structure and terrain characteristics (Curcio et al., 2022).

4.2 APPLICATION OF UAVS FOR SECURITY AND SURVEILLANCE

Security and surveillance are crucial aspects in various domains, including law enforcement, public safety, border control, critical infrastructure protection, and disaster management. UAVs, also known as drones, have emerged as valuable tools for enhancing security and surveillance capabilities. UAVs offer several advantages such as aerial perspective, real-time monitoring, and cost-effectiveness, making them an ideal choice for surveillance operations.

One of the key methods employed in security and surveillance is camera observation. UAVs equipped with high-resolution cameras can capture real-time video footage and images from vantage points that would otherwise be difficult or inaccessible for traditional surveillance methods. The aerial perspective provided by UAVs allows for a wider field of view, enabling operators to monitor large areas effectively. This capability is particularly valuable in situations where situational awareness, threat detection, and response times are critical (Husman et al., 2021).

Additionally, UAVs can be equipped with various types of sensors and technologies to enhance surveillance capabilities. For example, GPS tracking systems can be integrated into UAVs, enabling real-time tracking and monitoring of objects or individuals. This is particularly useful in law enforcement operations, border control, and search and rescue missions. Data mining and profiling techniques can also be employed to analyse the collected surveillance data, identify patterns, and generate actionable intelligence (Shakhatreh et al., 2019).

Strategic surveillance operations can benefit from the use of UAVs for covert or extended surveillance missions. UAVs can be deployed for stake-outs, allowing law enforcement agencies to discreetly monitor suspicious activities or individuals without arousing suspicion. The ability of UAVs to hover or fly at low altitudes enables them to maintain a covert presence, gathering valuable intelligence without compromising the operation (Jeelani and Gheisari, 2021).

Another significant application of UAVs in security and surveillance is biometric surveillance. UAVs can be equipped with facial recognition systems, thermal cameras, or other biometric sensors to identify individuals of interest or detect anomalies in crowds. This technology can be valuable in crowd management, border control, and event security, enabling authorities to quickly identify potential threats or persons of interest (Gohari et al., 2022).

The utilization of UAVs for security and surveillance operations offers several advantages. Firstly, UAVs provide rapid deployment capabilities, allowing for immediate response to incidents or emergencies. They can quickly reach remote or inaccessible areas, providing real-time situational awareness and aiding in decision-making processes. Secondly, UAVs are cost-effective compared to traditional manned aircraft or ground-based surveillance methods. They require fewer resources, such as personnel and infrastructure, and can cover larger areas in less time (Hildmann and Kovacs, 2019).

Various studies and practical implementations have highlighted the effectiveness of UAVs in security and surveillance applications. For instance, in law enforcement operations, UAVs have been utilized for monitoring public events, traffic surveillance, and crime scene investigations. The Royal Canadian Mounted Police has employed UAVs for search and rescue missions, accident reconstruction, and tactical operations. These examples demonstrate the versatility and effectiveness of UAVs in security and surveillance domains (Lagkas et al., 2018).

4.3 APPLICATION OF UAVS FOR CROWD/TRAFFIC MANAGEMENT

UAVs have emerged as valuable tools for crowd and traffic management, providing aerial assistance to ground control systems. With their mobility, agility, and ability to cover large areas, UAVs overcome the limitations of traditional fixed camera surveillance and offer unique advantages in managing crowds and traffic. One of the key applications of UAVs in crowd management is their ability to provide an aerial perspective. By flying at different altitudes, UAVs can capture real-time video footage and images of crowds in various locations and angles. This aerial view enables operators to monitor crowd movements, identify potential bottlenecks, and assess the overall crowd dynamics. It helps authorities make informed decisions and take appropriate actions to ensure public safety and smooth crowd flow (Mechan et al., 2023).

UAVs equipped with high-resolution cameras and advanced imaging technologies can provide detailed situational awareness in crowded areas. They can quickly detect overcrowding, identify potential safety hazards, and monitor compliance with crowd management protocols. This real-time monitoring capability allows for proactive crowd control measures, such as adjusting entry or exit points, redirecting crowd flow, or deploying additional resources as needed (Al-Sheary et al., 2017).

In traffic management, UAVs play a crucial role in monitoring road conditions, traffic congestion, and in identifying traffic violations. By capturing aerial views of road networks, intersections, and highways, UAVs provide valuable information on traffic patterns, bottlenecks, and areas requiring intervention. This information helps authorities optimize traffic flow, improve response times, and implement effective traffic management strategies.

UAVs are particularly useful in situations where ground-based surveillance systems face limitations. Fixed cameras have restricted fields of view and can only monitor specific areas. In contrast, UAVs can navigate through the airspace, change altitude and position, and cover a larger area efficiently. This flexibility allows for rapid deployment to areas of interest, providing real-time monitoring and enhancing overall situational awareness (Ramachandran et. al, 2021).

4.4 APPLICATION OF UAVS FOR PRECISION AGRICULTURE

With increasing population, urbanization, climatic changes, a degraded environment, changes in food preference, and the need for increased productivity, efficient methods to prevent pathogens and to grow healthy crops is crucial in the agriculture sector. UAVs can aid in this process and compensate for the reduced manpower. The precision agriculture method helps with reliable detection, accurate identification and proper quantification of pathogens, and other factors that affect the productivity of the crops (Faical et al., 2014).

4.5 IF A DRONE IS DESIGNED FOR LIDAR MAPPING AND PHOTOGRAMMETRY PURPOSES, THEN THE FUNCTIONAL REQUIREMENTS INCLUDE (EVERAERTS (2008):

- Low-altitude flight to obtain detailed images
- Flight altitude must vary based on subject height instead of considering ground as the reference point
- High speed may result in blurry images; hence, the optimum speed must be set based on the wind speed and climatic conditions
- UAVs must be capable of analysing the landscape, light intensity, and climatic conditions
- UAVs should have high endurance and range to cover the required area
- Capacity to capture high-resolution and multiple images
- LiDAR sensors must be able to measure the height of the points in the landscape

4.6 IF A DRONE IS DESIGNED FOR LIDAR MAPPING AND PHOTOGRAMMETRY PURPOSES (COLOMINA, 2014), THEN THE DESIGN REQUIREMENTS INCLUDE:

- Structural design must contribute to low aerodynamic drag
- Payload capability to carry camera and LiDAR sensor
- Preferably vertical take-off and landing capabilities so any ambiguity of images taken during horizontal and vertical take-off and landing can be reduced (Yasir et al., 2018)
- Flight controller to trace pre-planned flight path (Peng et al., 2011)
- Fixed-wing configuration which has high endurance when compared to multi-rotor configuration
- Coverage area of approximately 5 km^2–10 km^2 in one-time battery charge
- Camera used for photogrammetry must have pixel size of 1–3 cm and should be capable of capturing a minimum of one photo every two seconds
- Flight altitude should be minimum of 120 m above the ground.

4.7 IF DRONES ARE DESIGNED FOR PRECISION AGRICULTURE, THEN THE FUNCTIONAL REQUIREMENTS INCLUDE:

- The UAV must be capable of controlling its flight attitude based on the observation ground.
- It must be semi-autonomous; it should be capable of flying according to the pre-planned flight path using waypoints and attitude data.
- The UAV must be capable of sensing and avoiding obstacles during flight.
- The images obtained from the UAV need to be stabilized with the help of a gimbal.
- The UAV must be capable of sensing the environmental conditions, linking wind speed, light intensity, and so forth, and varying the camera settings accordingly (Delavarpour et al., 2021).
- The range and flight attitude are dependent on the agriculture field selected.
- If crop spraying function need to be performed by the UAV, it must be capable of carrying the additional payload like pesticide and fertilizers (Restas et al., 2021).
- It must be capable of hovering at a required spot while crop spraying (Restas, 2021).
- The cameras and data processing system must be capable of calculating certain characteristics like the chlorophyll content of the crops, the leaf water content, ground cover and leaf area index, and normalized difference vegetation index.

4.8 IF DRONES ARE DESIGNED FOR PRECISION FARMING, THEN THE DESIGN REQUIREMENTS INCLUDE:

- A multi-rotor UAV can be used if functions like crop spraying are performed (because of its hovering capability). If the function of the UAV is to monitor the crop and process the obtained data, a fixed wing drone can be used (Ashraf et al., 2018).
- UAVs used are required to fly at a low attitude and must have a long range and endurance.
- The UAV must be embedded with an RGB (Red Green Blue) camera, a thermal sensing camera, a multispectral camera, and a LiDAR system (Xu et al., 2020).
- A Global Navigation Satellite System (GNSS) must be present to follow the waypoints and to maintain the attitude.

- The flight controller must be capable of avoiding collisions (Yasin et al., 2020).
- A portable or mobile ground control station is required to perform the mission.

4.9 IF DRONES ARE DESIGNED FOR SECURITY AND SURVEILLANCE, THEN THE FUNCTIONAL REQUIREMENTS INCLUDE:

- The UAV must be capable of flying at low attitude and must have a high endurance.
- It must be capable of object tracking and locking a particular target when required.
- The target must remain locked even when the aircraft's platform is moving.
- The UAV must be capable of sensing humans or animals even in the absence of light or in the presence of a dense ground cover.
- Data collected by the UAV must be encrypted and must be resistant to hacking from other systems.
- Alerting signals or an alarm system needs to be present during emergency situations.
- During an event of a propeller failure crash scenario, the drone locations must be sent to the ground station to retrieve it.

4.10 IF DRONES ARE DESIGNED FOR SECURITY AND SURVEILLANCE, THEN THE DESIGN REQUIREMENTS INCLUDE:

- A fixed-wing drone configuration can be employed when the area under surveillance is huge; a multi-rotor UAV can be used for surveillance of small areas (Ashraf et al., 2018).
- The flight controller must be capable of avoiding collisions and obstacles (Steven et al., 2016) in the flying path of the UAV.
- The UAV must be capable of carrying thermal sensing cameras and a LiDAR sensor as payload.
- The ground control station needs to be mobile or portable for surveillance missions. If the surveillance area is predefined, then a fixed ground control station can be used.
- Crashworthiness improvement methods like on-board parachute deployment systems need to be present to prevent damage to the drone in the event of external attack or propeller failure.

4.11 IF DRONES ARE DESIGNED FOR CROWD CONTROL AND TRAFFIC MANAGEMENT, THEN THE FUNCTIONAL REQUIREMENTS INCLUDE:

- UAVs must fly at a low attitude, must have high endurance, and should be capable of hovering at a place when required.
- It must be capable of providing real transmission of images and videos captured by the camera.
- Image quality must be adjusted based on the environmental conditions.
- UAVs used for crowd management must be capable of analysing medical emergencies and should send a prompt for medical aid to on-ground officials.
- It must prevent high-impact drops or collisions into the crowd in the event of propeller failure.
- UAVs used for traffic management must be capable of identifying or tracking individual vehicles. The cameras used must be capable of reading the number plate information. In the event of accidents, the UAV must alert the on-ground officials to provide assistance.

4.12 IF DRONES ARE DESIGNED FOR CROWD CONTROL AND TRAFFIC MANAGEMENT, THEN THE DESIGN REQUIREMENTS INCLUDE:

- A multi-rotor configuration is required to provide hovering capabilities.
- The on-ground control station can be fixed at a particular spot.
- The data link must be capable of linking beyond visual range distances.
- The UAV structure needs to be capable of carrying the camera payload.
- The UAV structure must be crashworthy.
- Gimbal needs to be installed to stabilize the captured images.
- A fail-safe system and "detect and avoid modules" need to be present to regain control of the drone and assist in its landing during an event of failure and also to prevent collision with other flying modules (Steven, 2016).

4.13 UAV FLIGHT RULES AND REGULATIONS

Drone piloting has specific rules and regulation in every country just like passenger planes or fighter jets. This chapter discusses the rules and regulations set concerning authority and certain locations that are considered as no-fly zones.

The rules and regulations are dictated by the Ministry of Civil Aviation. The Director General of Civil Aviation (DGCA) released the regulations on 28 August 2018, to be effective from 1 December of the same year. Before that, there were no restrictions. This act was issued under the provision of Rule 15A and Rule 133A of the aircraft rules, 1937. According to this regulation, few restrictions on drone piloting are applied.

Drone classifications based on weight, nano, micro, small, medium, and large, are considered as categories of remotely piloted aircrafts.

Registration is one of the regulations. Just like other vehicles on the ground, drones have to be registered and have a Unique Identification Number (UIN), and an Unmanned Aircraft Operator Permit (UAOP) from the DGCA. Nano vehicles are exempt from this; they are not required to be registered. Imported vehicles have to obtain Equipment Type Approval (ETA) from the Wireless Planning and Coordination (WPC) wing, Department of Telecommunication. This also does not apply to nano category drones. Every UAV with a UIN has to have insurance, just like ground vehicles.

A pilot has to have proper training to fly drones (except nano and micro). The pilot has to be 18 years old, 10th pass, with good English communication skills to take the training programmes. Only training certificates from an organization approved by the DGCA and theory subjects prescribed by them are considered.

To fly drones in India, the pilot has to be Indian. No foreigner can fly drone in India. They have to hire an Indian or an Indian company with a drone piloting license if that is necessary, and then the UIN can be obtained.

The drone manufacturer has to install systems in the vehicle that include an identification plate, a flight controller with a logging system of flight data, a no permission no take-off system (NPNT), GPS, a return-to-home system, and an anti-collision light (Jawad, 2020; Steven et al., 2016). The minimum manufacturing standard provided by the DGCA has to be maintained, except for nano drones (Peng et al., 2011).

Drones cannot be flown in any random airspace. There are three types of zones for drones: red, yellow, and green zone depending on the flying restrictions. They are not allowed to be flown in red zones, which are near international borders, military bases, or around airports. Flying near parliament, Rastrapati Bhavan, state secretariat buildings in every state capital, Vijay Chowk in New Delhi, and strategic locations is also restricted. Flying over eco-sensitive zones such as national parks, wild life sanctuaries is also restricted. Moreover, flying over temporarily prohibited, dangerous areas is prohibited. Drones can be flown in yellow zones by filing a flight plan and an air defence clearance certificate. A green zone is completely uncontrolled airspace.

The pilot has to receive permission every time to fly the drone. Nano and micro drones can be flown below 50 feet and 200 feet respectively, without permission. Flying any type of civilian/commercial drone above 400 feet is restricted. Applications for flying drones can be applied for through the mobile app Digital Sky, which is the platform that gives permission as long as it is in the green zone. In other zones, permission will not be provided, and the NPNT system will prevent drones from taking off.

Drones cannot be flown at night. During the day, they cannot be flown away from the pilot's line of sight. The pilot has to be able to see the drone every minute while flying. They cannot be flown during rain, hail or snow, or thunderstorm activity. Flying from a mobile platform, such as moving vehicles, ships, or aircraft, is restricted.

Every event of a crash or a missing drone has to be reported to the DGCA and Bureau of Civil Aviation Security (BCAS) for purposes of security. Moreover, any drone with a UIN cannot be disposed of without notifying or receiving permission from the DGCA.

According to the DGCA, certain regions are classified as no-fly zones. These are as follows:

- Within a distance of 5 km from the perimeter of airports at Mumbai, Delhi, Chennai, Kolkata, Bangalore, and Hyderabad
- Within a distance of 3 km from the perimeter of any civil, private, or defence airport
- Above the Obstacle Limitation Surfaces (OLS) or PANS-OPS surfaces, whichever is lower, of an operational aerodrome, specified in Ministry of Civil Aviation (Height Restrictions for Safeguarding of Aircraft Operations) Rules, 2015 notified through *Gazette of India* notification GSR751(E) as amended from time to time
- Within permanent or temporary Prohibited, Restricted, and Danger Areas, including Temporary Reserved Airspace (TRA) and Temporarily Segregated Areas (TSA), as notified in the Aeronautical Information Publication (AIP), flights are restricted or prohibited, depending on the specific area type.
- Within 25 km from an international border, which includes Line of Control (LoC), Line of Actual Control (LAC), and Actual Ground Position Line (AGPL)
- Beyond 500 m (horizontal) into the sea from the coastline provided the location of ground station is on a fixed platform over land
- Within 3 km from the perimeter of military installations/facilities/where military activities/exercises are being carried out unless clearance is obtained from the local military installation/facility
- Within a 5 km radius from Vijay Chowk in New Delhi. However, this is subject to any additional conditions/restrictions imposed by local law enforcement agencies/authorities in view of the security
- Within 2 km from the perimeter of strategic locations/vital installations notified by the Ministry of Home Affairs (MHA) unless clearance is obtained from MHA

- Within 3 km from the radius of the State Secretariat Complex in state capitals
- From a mobile platform such as a moving vehicle, ship, or aircraft
- Over eco-sensitive zones around national parks and wildlife sanctuaries notified by the Ministry of Environment, Forests and Climate Change without prior permission.

4.14 LIST OF MATERIALS REQUIRED FOR ASSEMBLING A MINI VTOL CONFIGURATION FIXED WING DRONE

The list of components required to assemble a mini VTOL with fixed wing configuration is enlisted in Table 4.1. Also enlisted in the table are the number of components required and approximate cost of the component in Indian rupees.

Table 4.1 Bill of materials

Product number	*Component*	*Description*	*No. of items required*	*Price per item (in rupees)*
1	Motor	Brushless DC motor (12 kV)	5	500
2	Propeller	10 inch	5	50
3	ESC	Electronic speed controller (30 A)	5	350
4	Servo motor	To position control surface	4	100
5	PDB	Power distribution board	1	50
6	Battery	Li-Po battery (5,000 mAh)	1	3,000
7	Autopilot	GPS, Telemetry and sensors	1	15,000
8	Radio controller	To control UAV	1	4,000
9	Carbon fibre	To build the structure along with the wing	2 m^2	4,000/m^2
10	Plastic landing gears	To support the structure during landing and take-off	4	300
11	Aluminium frame	Wing main spar	1.5 m	200

Table 4.2 List of cameras

Make	*Model*	*Resolution*	*Lens Focal Length equiv)*	*Approx. Price (in Indian Rupees)*
Fujifilm	XF10	24 MP	28 mm	37,745
Ricoh	GR II	16.2 MP	28 mm	37,745
Canon	M100	24 MP	35 mm	50,000
Sony	Alpha a6000	24 MP	32 mm	57,000
Fujifilm	X-A5	24 MP	28 mm	68,000
Nikon	D3500	24 MP	32 mm	68,000
Canon	EOS Rebel SL2	24 MP	38 mm	52,000

Source: www.photomodeler.com/

4.15 LIST OF LOW-COST CAMERAS THAT CAN BE USED FOR PHOTOGRAMMETRY APPLICATIONS

There is a trend in assembling low-cost sensors in the body of the drones and achieve photogrammetric functions such as measurement of the height of the ground objects, generating the panoramic view of large area and locating objects on earth. Tabulated below (ref Table 4.2) is a list of low cost yet robust cameras with their approximate cost in Indian Rupees, focal length, resolution and model.

QUESTIONS AND ANSWERS

1. List 10 scientific applications of a UAV. List 10 societal applications of a UAV.

Answer: Some popular applications of UAVs mentioned in this chapter include LiDAR mapping, photogrammetry, precision agriculture, surveillance, crowd control and traffic measurement, disaster management, and archaeological applications. UAVs can also be used for search and rescue operations, wildlife monitoring, atmospheric research, and infrastructure inspection. In terms of societal applications, UAVs can be used for emergency response services such as delivering medical supplies or providing communication support during natural disasters. They can also be used for media coverage and entertainment purposes such as aerial photography or filming events. Additionally, UAVs can assist in environmental monitoring and conservation efforts.

2. State the specific requirements of a UAV for

a) Precision Agriculture, b) Photogrammetric measurements, c) Surveillance, d) Crowd control and traffic measurement, e) Disaster management, f) Archaeological applications.

Answer: a) Precision Agriculture: UAVs used for precision agriculture require high-resolution cameras, multispectral sensors, and GPS systems to capture data on crop health, soil moisture levels, and plant growth. They also need to be able to fly at low altitudes and have a long flight time.

b) Photogrammetric measurements: UAVs used for photogrammetric measurements require high-resolution cameras and GPS systems to capture accurate 3D models of terrain or structures. They also need to be able to fly at different altitudes and angles to capture the necessary data.
c) Surveillance: UAVs used for surveillance require high-resolution cameras with zoom capabilities, thermal imaging sensors, and real-time video transmission capabilities. They also need to be able to fly quietly and have a long flight time.
d) Crowd control and traffic measurement: UAVs used for crowd control and traffic measurement require high-resolution cameras with wide-angle lenses, real-time video transmission capabilities, and the ability to fly at low altitudes. They also need to be able to operate in urban environments with obstacles such as buildings or trees.
e) Disaster management: UAVs used for disaster management require high-resolution cameras with thermal imaging sensors, real-time video transmission capabilities, and the ability to fly in harsh weather conditions. They also need to be able to operate in areas with limited access or damaged infrastructure.
f) Archaeological applications: UAVs used for archaeological applications require high-resolution cameras with multispectral sensors, GPS systems, and the ability to fly at different altitudes. They also need to be able to capture data on large areas of land while avoiding damage or disturbance of historical sites.

3. What are the different requirements of a drone in civil applications? What are the requirement statements with respect to each of these applications? What are the payloads a drone has to be fitted with to achieve these requirements?

Answer: General requirements for drones in civil applications include compliance with airspace regulations, safety features such as obstacle avoidance systems, and reliable communication systems for real-time data transmission. The payloads needed for civil applications can vary depending on the specific use case, but may include high-resolution cameras, thermal imaging sensors, LiDAR sensors, multispectral sensors, and GPS systems.

4. How is a drone used for controlling crowd and manage traffic?

Answer: Drones can be used for crowd and traffic management by providing an aerial perspective that traditional fixed camera surveillance systems cannot offer.

By flying at different altitudes, drones can capture real-time video footage and images of crowds in various locations and angles. This aerial view enables operators to monitor crowd movements, identify potential bottlenecks, and assess the overall crowd dynamics. Drones can also be used to monitor traffic flow and identify areas of congestion or accidents. The real-time data captured by drones can be transmitted to ground control systems, allowing operators to make informed decisions and take appropriate actions to manage crowds or traffic.

5. Give examples how a drone can be used for medical applications.

Answer: Drones can be used for medical applications in various ways, including:

1. Delivering medical supplies: Drones can be used to deliver medical supplies such as vaccines, blood, or medications to remote or hard-to-reach areas. This can be especially useful in emergency situations where time is critical.
2. Transporting organs for transplant: Drones can be used to transport organs for transplant quickly and efficiently, reducing the time between organ retrieval and transplantation.
3. Monitoring public health: Drones equipped with sensors can be used to monitor air quality, water quality, and other environmental factors that affect public health.
4. Search and rescue operations: Drones equipped with thermal imaging sensors can be used to locate missing persons or injured hikers in remote areas.
5. Telemedicine: Drones equipped with cameras and communication systems can be used to provide telemedicine services to patients in remote areas who do not have access to healthcare facilities.

6. What are the flying rules for an UAV within civilian airspace?

Answer: The flying rules for UAVs in civilian airspace vary depending on the country and region. However, some common rules include:

1. Registering the drone with the relevant aviation authority.
2. Flying below a certain altitude, usually 400 feet or lower.
3. Keeping the drone within visual line of sight at all times.
4. Avoiding flying near airports or other restricted airspace.
5. Not flying over people or crowds.
6. Obtaining permission from property owners before flying over private property.
7. Following any additional regulations set by local authorities or aviation agencies.

It is important to note that failure to follow these rules can result in fines, legal action, and potential safety hazards for people and property on the ground.

7. Give a list of materials required for assembling a VTROL UAV.

Answer: Here is a list of materials required for assembling a VTOL UAV:

1. Airframe: This includes the main body of the UAV, wings, and other structural components.
2. Motors: Electric motors are typically used to power the propellers or rotors.
3. Propellers or rotors: These are used to generate lift and propulsion.
4. Flight controller: This is the brain of the UAV that controls its flight path and stability.
5. Battery: A high-capacity battery is needed to power the motors and other electronics on board.
6. Electronic speed controllers (ESCs): These control the speed of each motor.
7. Radio transmitter and receiver: These are used to control the UAV remotely.
8. GPS module: This provides location data for navigation and positioning.
9. Sensors: Various sensors such as accelerometers, gyroscopes, barometers, and magnetometers are used for stabilization, altitude hold, and other functions.
10. Camera or other payload equipment (optional): Depending on the intended use case, a camera or other payload equipment may be needed for data collection or other purposes.

8. Give technical specification for a quad copter. How does a quad copter differ from a fixed-wing UAV?

Answer:

1. Number of rotors: 4
2. Flight time: 15–30 minutes (depending on battery capacity and payload weight)
3. Maximum speed: 30–50 mph (depending on model)
4. Maximum altitude: 400 ft (as per Federal Aviation Administration regulations)
5. Payload capacity: Up to several pounds (depending on model)
6. Camera resolution: Up to 4K Ultra HD (depending on model)
7. Control range: Up to several miles (depending on model and regulatory restrictions)

A quadcopter differs from a fixed-wing UAV in several ways:

1. Design: A quadcopter has four rotors that provide lift and propulsion, while a fixed-wing UAV has wings that generate lift and require forward motion for propulsion.
2. Flight characteristics: A quadcopter can hover in place, fly in any direction, and perform agile manoeuvres such as flips and rolls, while a fixed-wing UAV requires forward motion to stay aloft and has limited manoeuvrability.
3. Payload capacity: Quadcopters typically have lower payload capacities than fixed-wing UAVs due to their design.
4. Flight time: Quadcopters generally have shorter flight times than fixed-wing UAVs due to the energy required to keep them aloft without forward motion.
5. Control range: Fixed-wing UAVs can typically fly longer distances than quadcopters due to their more efficient design, but may require more advanced control systems for stable flight at high speeds and altitudes.

9. Give a list of sensors and their specifications used as payload in different UAVs.

Answer: The sensors used as payload in different UAVs can vary depending on the specific application and mission requirements. However, here are some common types of sensors used in UAV payloads and their specifications:

1. Cameras: Cameras are one of the most common types of sensors used in UAVs for aerial photography, videography, mapping, and surveying. Camera specifications can vary widely depending on the intended use case, but may include resolution (e.g. 4K Ultra HD), frame rate, lens type, and image stabilization.
2. LiDAR: LiDAR (Light Detection and Ranging) sensors use laser pulses to create 3D maps of terrain or objects on the ground. LiDAR specifications may include range (e.g. up to 100 meters), accuracy (e.g. sub-centimeter), and point density.
3. Thermal imaging: Thermal imaging sensors detect heat signatures to create images that can be used for applications such as search and rescue, wildlife monitoring, or building inspections. Thermal imaging sensor specifications may include resolution (e.g. up to 640 x 480 pixels), temperature range (e.g. –40°C to +550°C), and sensitivity.
4. Multispectral/hyperspectral imaging: These sensors capture images across multiple spectral bands to provide detailed information about vegetation health, soil moisture content, or other environmental factors. Specifications may include spectral range (e.g. visible/near-infrared or

beyond), spatial resolution (e.g. down to 5 cm/pixel), and number of bands.

5. Gas/chemical sensors: These sensors detect specific gases or chemicals in the air for applications such as environmental monitoring or leak detection. Specifications may include detection range, sensitivity, and response time.
6. Magnetometers: Magnetometers measure magnetic fields for applications such as mineral exploration or archaeological surveys. Specifications may include sensitivity (e.g. down to 0.1 nT) and range.
7. Radar: Radar sensors use radio waves to detect objects on the ground or in the air for applications such as terrain mapping or obstacle avoidance. Specifications may include range (e.g. up to several kilometres), resolution, and frequency.

REFERENCES

Abd Rahman, Y.A. et al. (2018). Design and Fabrication of Small Vertical-Take-Off-Landing Unmanned Aerial Vehicle. *MATEC Web of Conferences* 152, 02023.

Al-Sheary, A., Almagbile, A. (2017). Crowd Monitoring System Using Unmanned Aerial Vehicle (UAV). *Journal of Civil Engineering and Architecture* 11. 10.17265/1934-7359/2017.11.004

Colomina, P.M. (2014). Unmanned Aerial System for Photogrammetry and Remote Sensing: A Review. *ISPRS Journal of Photogrammetry and Remote Sensing* 92, 79–97.

Curcio, A.C., Peralta, G., Aranda, M., Barbero, L. (2022). Evaluating the Performance of High Spatial Resolution UAV-Photogrammetry and UAV-LiDAR for Salt Marshes: The Cádiz Bay Study Case. *Remote Sensing* 14(15), 3582. https://doi.org/10.3390/rs14153582

Debnath, S., Paul, M., Debnath, T. (2023). Applications of LiDAR in Agriculture and Future Research Directions. *Journal of Imaging* 9(3), 57. https://doi.org/10.3390/jimaging9030057

Delavarpour, N., Koparan, C., Nowatzki, J., Bajwa, S., Sun, X. (2021). A Technical Study on UAV Characteristics for Precision Agriculture Applications and Associated Practical Challenges. *Remote Sensing* 13(6), 1204. https://doi.org/10.3390/rs13061204

Everaerts, J. (2008). The Use of Unmanned Aerial Vehicles (UAVs) for Remote Sensing and Mapping. *The International Archives of the Photogrammetry, Remote Sensing and Spatial Information Sciences* 37.

Faiçal, B.S., Pessin, G., Filho, G.P.R., Carvalho, A.C.P.L.F., Furquim, G., Ueyama, J. (2014). Fine-Tuning of UAV Control Rules for Spraying Pesticides on Crop Fields. *2014 IEEE 26th International Conference on Tools with Artificial Intelligence*, pp. 527–533. 10.1109/ICTAI.2014.85

Gaurav Singhal, Babankumar Barrod, Lini Mathew (2018), "Unmanned Aerial Vehicle Classification, Applications and challenger: A Review", (www.preprint.org)

Gohari, A., Ahmad, A., Rahim, R., Supa'at, A.S.M., Razak, S., Gismalla, M. (2022). Involvement of Surveillance Drones in Smart Cities: A Systematic Review. *IEEE Access* 10. 10.1109/ACCESS.2022.3177904

Hildmann, H., Kovacs, E. (2019). Review: Using Unmanned Aerial Vehicles (UAVs) as Mobile Sensing Platforms (MSPs) for Disaster Response, Civil Security and Public Safety. *Drones* 3(3), 59. https://doi.org/10.3390/drones3030059

Husman, M.A., Albattah, W., Abidin, Z.Z., Mustafah, Y.M., Kadir, K., Habib, S., Islam, M., Khan, S. (2021). Unmanned Aerial Vehicles for Crowd Monitoring and Analysis. *Electronics* 10(23), 2974. https://doi.org/10.3390/electronics10232974

Jeelani, I., Gheisari, M. (2021). Safety Challenges of UAV Integration in Construction: Conceptual Analysis and Future Research Roadmap. *Safety Science* 144, 105473, ISSN 0925-7535.

Lagkas, T., Argyriou, V., Bibi, S., Sarigiannidis, P. (2018). UAV IoT Framework Views and Challenges: Towards Protecting Drones as "Things". *Sensors* 18(11), 4015. https://doi.org/10.3390/s18114015

Mechan, F., Bartonicek, Z., Malone, D. et al. (2023). Unmanned Aerial Vehicles for Surveillance and Control of Vectors of Malaria and Other Vector-Borne Diseases. *Malar Journal* 22, 23. https://doi.org/10.1186/s12936-022-04414-0

Nex, F., Armenakis, C., Cramer, M., Cucci, D.A., Gerke, M., Honkavaara, E., Kukko, A., Persello, C., Skaloud, J. (2022). UAV in the Advent of the Twenties: Where We Stand and What Is Next. *ISPRS Journal of Photogrammetry and Remote Sensing* 184, 215–242, ISSN 0924-2716.

Peng, Z., Jikai, L. (2011). On New UAV Flight Control System Based on Kalman & PID. 2011 2nd International Conference on Intelligent Control and Information Processing, pp. 819–823. 10.1109/ICICIP.2011.6008362

Ramachandran, A., Sangaiah, A.K. (2021). A Review on Object Detection in Unmanned Aerial Vehicle Surveillance. *International Journal of Cognitive Computing in Engineering* 2, 215–228. ISSN 2666-3074.

Restas, A., Szalkai, I., Gyula, Ó. (2021). Drone Application for Spraying Disinfection Liquid Fighting against the COVID-19 Pandemic—Examining Drone-Related Parameters Influencing Effectiveness. *Drones* 5, 58. 10.3390/drones5030058

Roelofsen, S., Martinoli, A., Gillet, D. (2016, December). 3D Collision Avoidance Algorithm for Unmanned Aerial Vehicles with limited Field of View Constraints. *2016 IEEE 55th Conference on Decision and Control (CDC)*, Las Vegas, NV, USA, 2016, pp. 2555–2560, doi: 10.1109/CDC.2016.7798647.

Shakhatreh, H., Sawalmeh, A., Al-Fuqaha, A., Dou, Z., Almaitta, E., Khalil, I., Othman, N., Khreishah, A., Guizani, M. (2019). Unmanned Aerial Vehicles (UAVs): A Survey on Civil Applications and Key Research Challenges. *IEEE Access* 7, 48572–48634. 10.1109/ACCESS.2019.2909530

Xu, Z., Shen, X., Cao, L., Coops, N., Goodbody, T., Zhong, T., Zhao, W., Sun, Q., Ba, S., Zhang, Z., Wu, X. (2020). Tree Species Classification Using UAS-Based Digital Aerial Photogrammetry Point Clouds and Multispectral Imageries in Subtropical Natural Forests. *International Journal of Applied Earth Observation and Geoinformation* 92, 102173. 10.1016/j.jag.2020.102173

Yasin, J.N. et al. (2020). Unmanned aerial Vehicles (UAVs): Collision Avoidance Systems and Approaches. *IEEE*. 10.1109/Access 2020.3000064

Chapter 5

Classifications of different UAVs

OVERVIEW

Different Unmanned Aerial Vehicle (UAV) classification schemes have been proposed to help differentiate existing UAV systems based on their operational characteristics and their capabilities (Gaurav, 2018). Several of these schemes are also of regulatory importance since the metrics used directly correlate with risk from ground impact or mid-air collision accidents (Yasin, 2020), (Mcfadyen, 2016). This chapter provides characteristic UAV classifications from a variety of sources, both civilian and military (Jyoti, 2020).

This chapter discusses various types of UAVs. The design and development of UAVs is majorly dependent on their application and mission requirements; hence, different parameters are considered for their classification (Aabid et al., 2022). This chapter consists of a brief explanation of different design parameters of UAVs and classifies them as per some of the design parameters.

Different design parameters of UAVs are based on maximum take-off weight, lift producing devices, propulsion system, landing type, applications, flight altitude, and level of autonomy. There are various other parameters that can be used, but the characteristics related to further topics of discussion were considered (Ahmed et al., 2022).

UAV is an acronym for Unmanned Aerial Vehicle, which is an aircraft with no pilot on board. UAVs can be remote-controlled aircraft (e.g. flown by a pilot at a ground control station), or can fly autonomously based on preprogrammed flight plans or more complex dynamic automation systems. UAVs are currently used for a number of missions, including reconnaissance and attack roles. For the purposes of this chapter, and to distinguish UAVs from missiles, a UAV is defined as being capable of controlled, sustained level flight and powered by a jet or reciprocating engine. In addition, a cruise missile can be considered to be a UAV, but is treated separately on the basis that the vehicle is the weapon. The acronym UAV has been expanded in some cases to UAVS (Unmanned Aircraft Vehicle System). The Federal Aviation Administration (FAA) has adopted the acronym UAS (Unmanned Aircraft System), to reflect the fact that these complex systems include ground stations and other elements besides the

DOI: 10.1201/9781003432685-5

actual air vehicles. Officially, the term "unmanned aerial vehicle" was changed to "unmanned aircraft system" to reflect the fact that these complex systems include ground stations and other elements besides the actual air vehicles. The term "UAS," however, is not widely used as the term "UAV" has become part of the modern lexicon. The military role of UAVs is growing at unprecedented rates. In 2005, tactical and theatre-level Unmanned Aircraft (UA) alone, had flown over 100,000 flight hours in support of Operation Enduring Freedom and Operation Iraqi Freedom. Rapid advances in technology are enabling more and more capability to be placed on smaller airframes, which is spurring a large increase in the number of Small Unmanned Aerial Systems (SUAS) being deployed on the battlefield. The use of UASs in combat is so new that no formal US Department of Defense (DoD)-wide reporting procedures have been established to track SUAS flight hours. As the capabilities grow for all types of UAVs, nations continue to subsidize their research and development, leading to further advances that enable them to perform a multitude of missions. UAVs no longer only perform Intelligence, Surveillance, and Reconnaissance missions, although this still remains their predominant type (Martinez, 2019). Their roles have expanded to areas including Electronic Attack (EA), strike missions, suppression and/or destruction of enemy air defence (SEAD/DEAD), network node or communications relay, Combat Search And Rescue (CSAR), and derivations of these themes. These UAVs range in cost from a few thousand dollars to tens of millions of dollars, and the aircraft used in these systems (Fares et.al, 2016).

Different UAV classification schemes have been proposed to help differentiate existing systems based on their operational characteristics and their capabilities (Gaurav, 2018). Several of these schemes are also of regulatory importance since the metrics used directly correlate with risk from ground impact or midair collision accidents (Yasin, 2020). This chapter provides characteristic UAV classifications from a variety of sources, both civilian and military (Ukaegbu et al., 2021).

This chapter discusses various types of UAVs. The design and development of UAVs are majorly dependent on their application and mission requirements; hence, different parameters are considered for their classification (Aabid et al., 2022). This chapter consists of a brief explanation of different design parameters of UAVs and classifies them as per some of the design parameters.

Classification of UAVs based on different design parameters discussed in this chapter are maximum take-off weight, lift producing devices, propulsion system, landing type, applications, flight altitude, level of autonomy and so on. There are various other secondary parameters of UAVs that are country specific or specific to some production groups which are also discussed in this chapter. (Ahmed et al., 2022)

5.1 CLASSIFICATION OF UNMANNED AERIAL VEHICLES (UAVS)

Classification of UAVs does not have one standard. They are classified according to various factors and characteristics of UAVs (Gaurav, 2018). Some of the key factors for classifying a UAV are its (a) size, (b) range, (c) endurance, (d) altitude and Maximum Take-Off Weight (MTOW), (e) aerodynamics, (f) landing system etc. Classification of UAVs not only helps in differentiating the UAVs used for different missions but also helps for regulatory purposes to impose rules for their operation based on the characteristics of the drone. Some the major classifications are discussed in the subsequent sections (Giordan et al., 2020).

5.2 A GENERIC CLASSIFICATION OF UAVS

A generic classification of UAVs based on the altitude of operation and their wing types are depicted in the hierarchy diagram. The second level of classification includes high- or low-flying altitude and fixed or rotary wing types. These subclassifications are further augmented with their operation capabilities (Figure 5.1).

5.3 CLASSIFICATION OF UAVS BASED ON MAXIMUM TAKE-OFF WEIGHT (MTOW) AND RANGE

MTOW is the maximum mass at which the UAV is certified to take off due to structural and other limits. Also, MTOW is an important specification which decides to the maximum range of operation of the UAV. The classification of UAV according to the MTOW and range is enlisted in Table 5.1.

5.4 CLASSIFICATION BASED ON PROPULSION SYSTEM USED IN UAVS

The selection of propulsion systems majorly depends on the task and the flight goals. It is necessary to consider the payload and structural capacity of the system. Different propulsion systems are used by different UAVs (Alulema et al., 2021), and their classification is as depicted in Figure 5.2.

5.5 CLASSIFICATION OF UAVS BASED ON LIFT PRODUCING DEVICES

The two major lift producing devices include fixed wing and rotary wing, and the selection of the device depends on various factors like applications, power

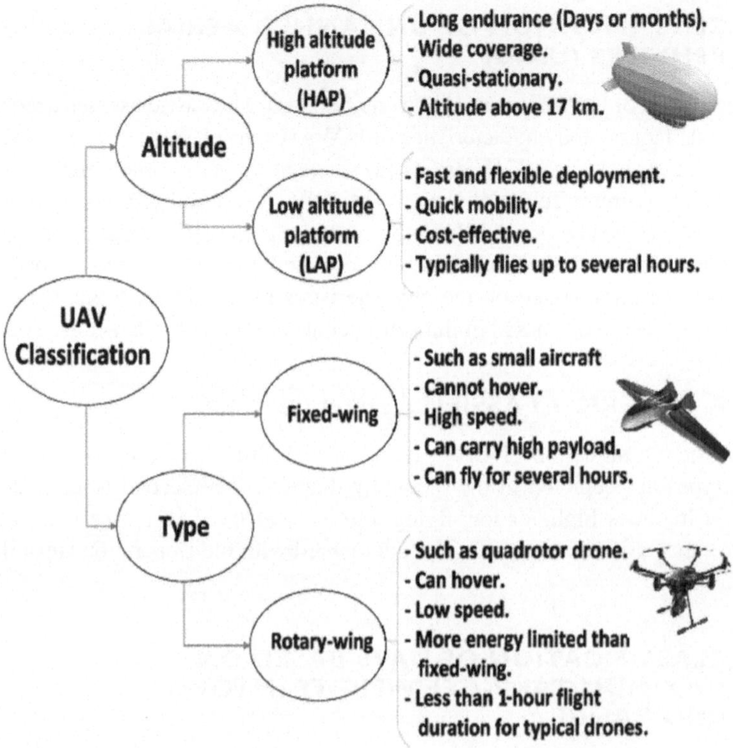

Figure 5.1 A generic classification of UAVs.

Table 5.1 Classification of UAVs based on MTOW

Type	*Maximum Take-Off Weight*	*Maximum Range*
Nano	200 grams	5 km
Micro	2 kg	25 km
Mini	20 kg	40 km
Light	50 kg	70 km
Small	150 kg	150 km
Tactical	600 kg	150 km
MALE	1,000 kg	200 km
HALE	1,000 kg	250 km
Heavy	2,000 kg	1,000 km
Super Heavy	2,500 kg	1,500 km

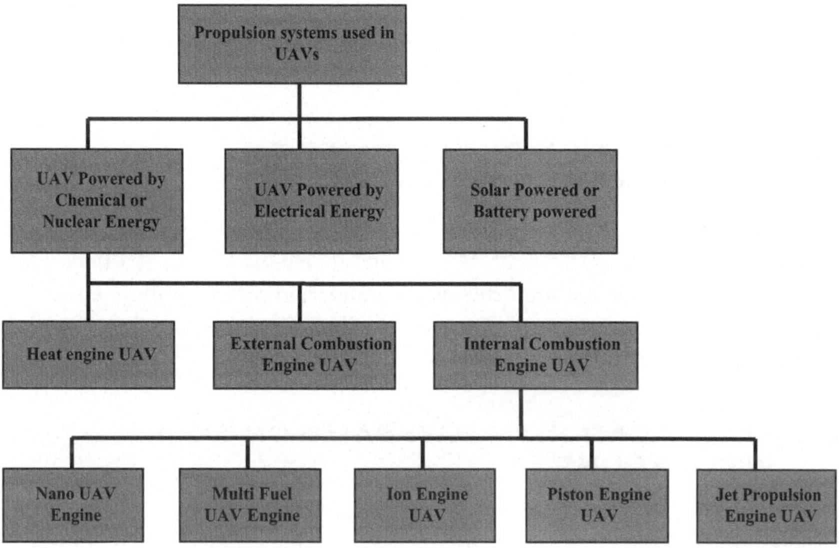

Figure 5.2 Hierarchical classification of UAVs based on propulsion system used.

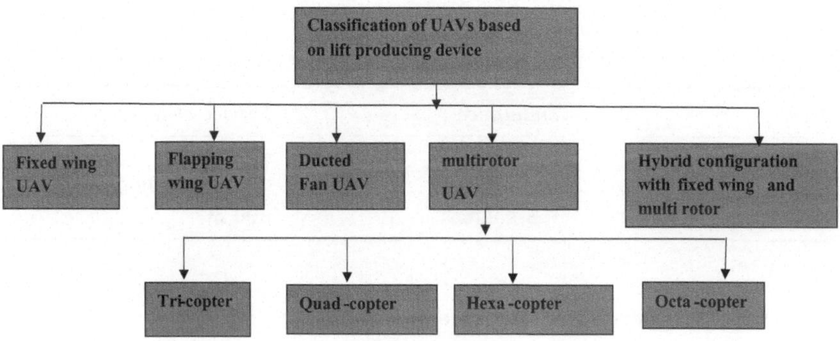

Figure 5.3 Classification of UAVs based on lift producing device.

source, range of operation, and so on. Further, the classification of UAVs based on the lift producing devices is depicted in Figure 5.3.

5.6 CLASSIFICATION OF UAVS BASED ON LANDING TYPE

Landing type is in turn related to the type of lifting device used in a UAV. Two major types of landing include Horizontal Take-Off and Landing (HTOL) and Vertical Take-Off and Landing (VTOL) (Roman, 2018). HTOL is most common in fixed-wing UAVs, while VTOL is widely used by single and multirotor UAVs. Hybrid configurations can perform both horizontal and vertical landing depending on their requirements (Ozdemir, 2014; Yasir, 2018).

5.7 CLASSIFICATION OF UAVS BASED ON ENDURANCE AND SERVICE CEILING

Endurance of flight operation of a UAV and its service ceiling depends on many factors such as its fuel capacity, MTOW and its payload capacity (Table 5.2). Based on the mission requirements of endurance and service ceiling, a suitable UAV is selected. These two factors can also help in estimating the refuelling time, which in turn aids in preparing the flight plan (Reid et al., 2021).

5.8 CLASSIFICATION OF UAVS BASED ON THEIR APPLICATIONS

Since the UAV is a multi-utility vehicle, it has a wide range of applications. Therefore, application-specific UAVs are designed and developed by different production agencies the world over for specific usage and client applications. The broad classifications of UAVs based on applications of different segments are as tabulated in Table 5.3 (Singhal, 2018):

Table 5.2 Classification of UAVs based on endurance and service ceiling

Type	*Endurance*	*Service ceiling*
Long	*≥24 hours*	20,000 m
Medium	*5–24 hours*	7,000 m–8,000 m
Low	*< 5 hours*	300 m

Table 5.3 Classification of UAV based on its applications

Civil	*Environment*	*Defence*
Aerial Photography	Air quality monitoring	High-endurance combat vehicle
Surveying of Construction	Soil Monitoring & Survey	War zone medical supply
Mining	Crop monitoring	Spying of chemicals
Aerial Delivery	Water inspection	Border surveillance
Logistics Supply	Underwater inspection	Bomb dropping
Precision Agriculture	Pollution monitoring	Aerial launching of missile
Disaster Management		
Surveillance		

5.9 CLASSIFICATION OF UAVS BASED ON LEVEL OF AUTONOMY

Autonomy is a complex concept, especially when artificial intelligence (AI)-based systems are catching up with human intelligence and are looking to go further. It applies to machines as much as it applies to humans. Hence, at one level it is a topic of psychological studies. Even in its engineering interpretations, it may well assume many hues. But one thing that has to be kept in the mind is human control over the extent of autonomy in a system. Responding to the fast-paced developments in the domain of self-driving cars, the automotive industry has come out with a standard of its own comprising six levels of autonomy. Table 5.4 lists some of these approaches to define levels of autonomy in AI-enabled autonomous systems.

Based on the range of flying altitude, the maximum gross weight of a UAV, and speed of its operations, the US DoD has classified UAVs into five categories, which are summarized in Table 5.5.

5.10 US DEPARTMENT OF DEFENSE CLASSIFICATION OF UNMANNED AERIAL SYSTEMS

US DoD takes into consideration many factors to classify an Aerial Unmanned System (AUS) of which UAV. It takes into account the Maximum Gross Take off Weight (MGTOW), normal operating altitude, speed of flying to classify a AUS into five groups which is tabulated in Table 5.5.

Table 5.4 Classification of UAVs based on autonomy

Autonomous control levels	*Level descriptor*
0	Remotely piloted vehicle
1	Execute pre-planned mission
2	Changeable mission
3	Robust response to real-time faults/events
4	Fault/event adaptive vehicle
5	Real-time multi-vehicle coordination
6	Real-time multi-vehicle cooperation
7	Battle space knowledge
8	Battle space cognizance
9	Battle space swarm cognizance

Table 5.5 US DoD classification of UAS

U.S. DoD UAS Group Descriptions				
UAS Groups	*MGTOW (Kg)*	*Normal Operating Altitude (m)*	*Speed (km/hr)*	*Representative UAS*
Group-1	0–10	< 366 AGL*	185	Raven (RQ-11), WASP
Group-2	10–25	< 1,067 AGL	463	Scan Eagle
Group-3	< 600	< FL 180 # (18,000 ft or 5,486 m)	463	Shadow (RQ-7B), Tier II / STUAS
Group-4	>600	< FL 180 (18,000 ft or 5,486 m)	Any Speed	Fire Scout (MQ-8B, RQ-8B), Predator (MQ-1A/B), Sky Warrior ERMP (MQ-1C)
Group-5	> 600	> FL 180 (18,000 ft or 5,486 m)	Any Speed	Reaper (MQ-9A), Global Hawk (RQ-4), BAMS (RQ-4N)

MGTOW - Maximum Gross Take-Off Weight
AGL - Above ground level
FL 180–18,000 ft above mean sea level when the pressure at sea level is 1013.2 milli bar

5.11 NATO CLASSIFICATION OF UAS

The NATO group of countries have taken into consideration following factors of an UAS to classify them into two classes and further sub categories. The factors taken for classification are Normal operating altitude above MSL/AGL, Normal mission radius, application it is employed for. The classification is depicted in the Table 5.6.

5.12 INDIAN CLASSIFICATION OF UAS

Indian industry has classified UASs into five categories depending upon the maximum take-off weight (MOTW). The nine operational parameters of these five categories are enumerated in the rows of the classification table. The operational parameters are

UIN[1]-Unique Identification Number; UAOP[2]: Unmanned Aerial Operators Permit;

NPNT[3]: No Permission No Take-off; AGL[4]: Above Ground Level; LOS[5]: Line Of Sight; ADC/FIC[6]: Air Defence Clearance/Flight Information Centre; IAUA[7]: Indoor and Uncontrolled Airspace; UA[8]: Uncontrolled Airspace;

Table 5.6 NATO classification of UAS

Class	*Category*	*Normal operating altitude*	*Normal mission radius*	*Example platform*	*Normal employment*
Class I (< 150 Kg)	Small (> 15 kg)	Up to 1.5 km AGL*	50 km (LoS[@])	Scan Eagle	Tactical unit
	Mini (< 15 kg)	Up to 914 m AGL	25 km (LoS)	Skylark	Tactical subunit (manual or hand launch)
	Micro (< 2 kg)	Up to 61 m AGL	5 km (LoS)	Black widow	Tactical subunit (manual or hand launch)
Class II (150 Kg to 600 Kg)	Tactical	Up to 5.5 km AGL	200 km (LoS)	Hermes450 Aerostar Ranger	Tactical formation
Class III (> 600 Kg)	Strike/ combat	Up to 19.8 km	Unlimited (BLoS[&])	Reapor	Strategic / National
	HALE[1]	Up to 19.8 km	Unlimited (BLoS)	Global Hawk	Strategic / National
	MALE[2]	Up to13.7 km MSL[#]	Unlimited (BLoS)	Predator B Predator A Heron	Operational / Theatre

AGL - Above Ground Level
MSL - Mean Sea Level
LoS - Line of Sight
BLoS - Beyond Line of Sight
HALE - High-Altitude Long-Endurance
MALE - Medium Altitude Long Endurance

Table 5.7 Indian classification of UAS

Parameters	*Nano*	*Micro*	*Small*	*Medium*	*Large*
Weight	≤ 250 g	250 g to 2 kg	2 kg to 25 kg	25 kg to 150 kg	• 150 kg
UIN[1]	No	Yes	Yes	Yes	Yes
UAOP[2]	No	No	Yes	Yes	Yes
NPNT[3]	No	Yes	Yes	Yes	Yes
Allowed height (AGL[4])	15 m	61 m	122 m	122 m	122 m
Night Operations	No	No	No	No	No
Visual LOS[5]	Yes	Yes	Yes	Yes	Yes
Flight Plan	No	No	Yes	Yes	Yes
ADC/FIC[6]	No	No	Yes	Yes	Yes
Flight Path Planning	No	Yes	Yes	Yes	Yes
Operation Envelope	IAUA[7]	UA[8]	CAUA[9]	CAUA	CAUA
Security Clearance	No	Yes	Yes	Yes	Yes
Pilot Training	No	No	Yes	Yes	Yes

UIN - Unique Identification Number; UAOP[2]: Unmanned Aerial Operators Permit; NPNT[3]: No Permission No Take-Off; AGL: Above Ground Level; LoS[5]: Line of Sight; ADC/FIC: Air Defence Clearance/Flight Information Centre; IAUA[7]: Indoor and Uncontrolled Airspace; UA[8]: Uncontrolled Airspace; CAUA[9]: Controlled and Uncontrolled Airspace.

CAUA[9]: Controlled and Uncontrolled Airspace. The Indian classification of UAS is tabulated in Table 5.7.

QUESTIONS AND ANSWERS

1. What are some different criteria used for classifying the family of autonomous vehicles systems?

Answer: The classification of autonomous vehicle systems can be based on various criteria, including:

Autonomy Levels: Autonomous vehicle systems are often classified based on their level of autonomy, as defined by the Society of Automotive Engineers (SAE) International. The SAE levels range from 0 to 5, where:

Level 0: No automation – The driver is in complete control of the vehicle.

Level 1: Driver assistance – The system can assist with specific functions, such as steering or acceleration, but the driver remains fully responsible.

Level 2: Partial automation – The system can control both steering and acceleration/deceleration simultaneously, but the driver must be ready to take control at any time.

Level 3: Conditional automation – The system can manage most aspects of the driving task under certain conditions, but the driver should be prepared to intervene when requested.

Level 4: High automation – The system can perform all driving tasks under specific conditions or within specific areas, with the driver as a fallback if needed.

Level 5: Full automation – The system can handle all driving tasks under all conditions, and no human intervention is required.

Functionality: Autonomous vehicle systems can also be classified based on their primary functionality. Some common classifications include:

Passenger vehicles: Autonomous systems designed for transporting passengers, such as self-driving cars or autonomous taxis.

Commercial vehicles: Autonomous systems used for transportation and logistics purposes, including autonomous trucks, delivery drones, or robotic shuttles.

Industrial vehicles: Autonomous systems designed for specific industrial applications, such as self-driving, forklifts, agricultural robots, or mining vehicles.

Application Domain: Another criterion for classification is the application domain or specific use case for autonomous vehicle systems. Examples include:

Urban driving: Systems optimized for navigating complex urban environments and handling traffic congestion.

Highway driving: Systems designed for long-distance travel on highways, focusing on maintaining lane position and managing higher speeds.

Off-road driving: Systems tailored for rugged terrain, such as self-driving vehicles used in agriculture, mining, or exploration.

Sensor Suite: Classification can also be based on the types and complexity of sensors used by autonomous vehicle systems. These may include cameras, radar, LiDAR, ultrasonic sensors, or other technologies.

2. What are some different criteria used for classifying UAVs and drones?

Answer: Different parameters are considered for the classification of UAVs and drones, such as their application and mission requirements. Some of the design parameters used for classification include propulsion systems, lift producing devices, Maximum Take-Off Weight (MTOW), range of operations, and level of autonomy.

3. Give a hierarchical diagram of classification of UAVs based on generic properties.

Answer: Figure 5.1

4. Classify UAVs based on MTOW and range of operations.

Ans: Based on the provided MTOW and range of operations, UAVs can be classified as follows:

1. Nano:
   Maximum Take-off Weight: Up to 200 grams
   Range of Operations: Up to 5 km
2. Micro:
   Maximum Take-off Weight: Up to 2 kg
   Range of Operations: Up to 25 km
3. Mini:
   Maximum Take-off Weight: Up to 20 kg
   Range of Operations: Up to 40 km
4. Light:
   Maximum Take-off Weight: Up to 50 kg
   Range of Operations: Up to 70 km
5. Small:
   Maximum Take-off Weight: Up to 150 kg
   Range of Operations: Up to 150 km
6. Tactical:
   Maximum Take-off Weight: Up to 600 kg
   Range of Operations: Up to 150 km
7. MALE (Medium Altitude Long Endurance):
   Maximum Take-off Weight: Up to 1,000 kg
   Range of Operations: Up to 200 km
8. HALE (High Altitude Long Endurance):
   Maximum Take-off Weight: Up to 1,000 kg
   Range of Operations: Up to 250 km
9. Heavy:
   Maximum Take-off Weight: Up to 2,000 kg
   Range of Operations: Up to 1,000 km
10. Super Heavy:
    Maximum Take-off Weight: Up to 2,500 kg
    Range of Operations: Up to 1,500 km

These classifications provide a general categorization based on the maximum take-off weight and the range of operations for each UAV type.

5. Classify UAVs based on the different propulsion systems employed

Answer:

a. UAV Powered by chemical or nuclear energy:
   Heat Engine UAV: Heat engine-powered UAVs utilize the energy produced by chemical reactions, such as combustion, to generate heat. This heat is then converted into mechanical energy, which propels the UAV. The heat engine can be powered by various fuels, including hydrocarbons or other combustible materials.

   External Combustion Engine UAV: External combustion engine-powered UAVs employ engines that generate power through an external heat source. These engines, such as steam engines or Stirling engines, use heat from a combustion process to operate and produce mechanical energy. The heat is transferred to a working fluid, which drives the engine and propels the UAV.

   Internal Combustion Engine UAV: Internal combustion engine-powered UAVs utilize the energy released by the combustion of fuel within the engine. This type of propulsion system is commonly found in traditional vehicles and aircraft. Internal combustion engines in UAVs can be further classified into various types:

   a. Nano UAV Engine: These engines are specifically designed for small, lightweight UAVs known as nano drones. They are compact, efficient, and optimized for the limited size and power requirements of nano UAVs.
   b. Multi-Fuel UAV Engine: Multi-fuel engines are capable of running on different types of fuels, offering flexibility in terms of fuel availability and usage. They can accommodate a range of fuels, including gasoline, diesel, or aviation fuel.
   c. Ion Engine UAV: Ion engines utilize electric fields to ionize and accelerate propellant gases, creating thrust. This type of engine is more commonly used in space exploration applications, but there are also UAVs that employ ion propulsion for extended flight durations and increased efficiency.
   d. Piston Engine UAV: Piston engines, also known as reciprocating engines, are commonly used in smaller UAVs. They use pistons to convert the pressure generated by the combustion process into rotational motion, which propels the UAV.
   e. Jet Propulsion Engine UAV: Jet engines are widely used in larger UAVs. They work on the principle of sucking in air, compressing it, and then combusting it with fuel to create high-velocity exhaust gases. The propulsion is achieved through the thrust generated by the high-speed exhaust gases.

b. UAV Powered by Electrical Energy:

UAVs powered by electrical energy use electric motors as their propulsion system. These motors are typically powered by rechargeable batteries. Electric-powered UAVs are known for their quiet operation, efficiency, and suitability for smaller drones used in photography, videography, and recreational purposes.

c. Solar Powered or Battery Powered:

Solar-powered UAVs use photovoltaic cells to convert sunlight into electrical energy. These cells are typically integrated into the wings or body of the drone, allowing it to harness solar energy and charge on-board batteries. Solar-powered UAVs are commonly used for long-endurance missions, environmental monitoring, and aerial surveying. Battery-powered UAVs, on the other hand, rely solely on rechargeable batteries for their electrical energy, providing a limited flight duration but offering versatility and ease of use.

6. Classify UAVs based on different lift producing devices

Answer:

a. Fixed-wing UAV: Fixed-wing UAVs are designed with a fixed-wing structure, similar to traditional airplanes. They generate lift through the shape of their wings and require forward motion to maintain flight. Fixed-wing UAVs are known for their efficiency, long-range capabilities, and stable flight. They are commonly used for aerial mapping, surveillance, and long-endurance missions.
b. Flapping Wing UAV: Flapping wing UAVs mimic the flight of birds or insects by flapping their wings. These drones achieve lift and propulsion by imitating the natural motion of wings. Flapping wing UAVs are still in the experimental stage and are being developed for specialized applications such as biomimicry research and micro-scale aerial robotics.
c. Ducted Fan UAV: Ducted fan UAVs feature a ducted fan system that encloses the propeller within a protective housing. This design enhances safety and efficiency by reducing the risk of injuries from exposed propellers and increasing thrust generation. Ducted fan UAVs are commonly used for indoor flight, close-quarters operations, and certain specialized applications.
d. Multirotor UAV: Multirotor UAVs, also known as rotorcraft UAVs, are characterized by multiple rotors that provide lift and propulsion. They are highly manoeuvrable and capable of Vertical Take-Off And Landing (VTOL). Multirotor UAVs can hover in place, change direction quickly,

and navigate in confined spaces. They are widely used for aerial photography, videography, inspection tasks, and recreational purposes.

e. Hybrid Configuration with Fixed Wing and Multirotor: Hybrid UAVs combine the features of fixed-wing and multirotor configurations. These drones are capable of VTOL like multirotors but can transition into forward flight like fixed-wing aircraft. Hybrid UAVs offer the benefits of both configurations, providing long-range capabilities, extended endurance, and the ability to hover and manoeuvre in tight spaces. They are often used for surveying, mapping, and surveillance missions.

Multirotor UAV types:

a. Tri-copter: Tri-copters are multirotor UAVs equipped with three rotors. They provide agility and manoeuvrability, but their stability and payload capacity are generally lower compared to quadcopters and other multirotor configurations.
b. Quadcopter: Quadcopters are the most common type of multirotor UAVs. They feature four rotors arranged in a square or X configuration. Quadcopters offer a good balance between stability, manoeuvrability, payload capacity, and flight time. They are used widely for various applications, including aerial photography, delivery services, and recreational flying.
c. Hexacopter: Hexacopters have six rotors, providing increased stability, redundancy, and payload capacity compared to quadcopters. The additional rotors enhance the drone's ability to withstand motor failures and continue flying safely. Hexacopters are often used for professional photography, videography, and industrial inspections.
d. Octocopter: Octocopters feature eight rotors, providing even greater redundancy, stability, and payload capacity. They are capable of carrying heavier cameras or equipment, and are commonly used in professional cinematography, heavy payload lifting, and industrial applications that require robust and reliable UAVs.

7. What are the criteria for classifying a UAV?
   a. based on NATO criteria
   b. based on Indian criteria

Answer:

a. Based on NATO criteria:
   1. Small UAS:
      - Category: Small (> 15 kg)
      - Normal operating altitude: Up to 1.5 km Above Ground Level (AGL)

- Normal mission radius: 50 km Line of Sight (LoS)
- Example platform: Scan Eagle
- Normal employment: Tactical unit

2. Class I UAS:
   - Category: Mini (< 150 kg)
   - Normal operating altitude: Up to 914 m AGL
   - Normal mission radius: 25 km LoS
   - Example platform: Skylark
   - Normal employment: Tactical subunit (manual or hand launch)
   - Category: Micro (< 2 kg)
   - Normal operating altitude: Up to 61 m AGL
   - Normal mission radius: 5 km LoS
   - Example platform: Black Widow
   - Normal employment: Tactical subunit (manual or hand launch)
3. Class II UAS:
   - Category: Tactical (150 kg to 600 kg)
   - Normal operating altitude: Up to 5.5 km AGL
   - Normal mission radius: 200 km LoS
   - Example platforms: Hermes 450, Aerostar, Ranger
   - Normal employment: Tactical formation
   - Category: Strike/Combat
   - Normal operating altitude: Up to 19.8 km
   - Normal mission radius: Unlimited Beyond Line of Sight (BLoS)
   - Example platform: Reaper
   - Normal employment: Strategic/National
4. Class III UAS:
   - Category: High-Altitude Long Endurance (HALE)
   - Normal operating altitude: Up to 19.8 km
   - Normal mission radius: Unlimited BLoS
   - Example platform: Global Hawk
   - Normal employment: Strategic/National
   - Category: Medium-Altitude Long Endurance (MALE)
   - Normal operating altitude: Up to 13.7 km Mean Sea Level (MSL)
   - Normal mission radius: Unlimited BLoS
   - Example platforms: Predator B, Predator A, Heron
   - Normal employment: Operational/Theatre

b. Based on Indian Criteria:
The Indian classification of Unmanned Aerial Systems (UAS) categorizes them into different types based on various parameters. Following is a description of each type:

1. Nano:
   Weight: ≤ 250 g
   UIN (Unique Identification Number): No
   UAOP (UAS Operator Permit): No
   NPNT (No Permission No Take-off): No
   Allowed Height (AGL): 15 m
   Night Operations: No
   Visual Line of Sight (LoS): Yes
   Flight Plan: No
   ADC/FIC (Air Defence Clearance/Flight Information Centre): No
   Flight Path Planning: No
   Operation Envelope: Identified Aerial Unmanned Aircraft (IAUA)
   Security Clearance: No
   Pilot Training: No

2. Micro:
   Weight: 250 g to 2 kg
   UIN: Yes
   UAOP: No
   NPNT: Yes
   Allowed Height (AGL): 61 m
   Night Operations: No
   Visual LOS: Yes
   Flight Plan: No
   ADC/FIC: No
   Flight Path Planning: Yes
   Operation Envelope: Unmanned Aircraft (UA)
   Security Clearance: Yes
   Pilot Training: No

3. Small:
   Weight: 2 kg to 25 kg
   UIN: Yes
   UAOP: Yes
   NPNT: Yes
   Allowed Height (AGL): 122 m
   Night Operations: No
   Visual LOS: Yes
   Flight Plan: Yes
   ADC/FIC: Yes
   Flight Path Planning: Yes
   Operation Envelope: Controlled Airspace Unmanned Aircraft (CAUA)
   Security Clearance: Yes
   Pilot Training: Yes

4. Medium:
   Weight: 25 kg to 150 kg
   UIN: Yes
   UAOP: Yes
   NPNT: Yes
   Allowed Height (AGL): 122 m
   Night Operations: No
   Visual LOS: Yes
   Flight Plan: Yes
   ADC/FIC: Yes
   Flight Path Planning: Yes
   Operation Envelope: Controlled Airspace Unmanned Aircraft (CAUA)
   Security Clearance: Yes
   Pilot Training: Yes

5. Large:
   Weight: > 150 kg
   UIN: Yes
   UAOP: Yes
   NPNT: Yes
   Allowed Height (AGL): 122 m
   Night Operations: No
   Visual LOS: Yes
   Flight Plan: Yes
   ADC/FIC: Yes
   Flight Path Planning: Yes
   Operation Envelope: Controlled Airspace Unmanned Aircraft (CAUA)
   Security Clearance: Yes
   Pilot Training: Yes

8. Give the Classification of UAVs based on Level of Autonomy

Answer: Based on the autonomy levels provided, UAVs can be classified into different categories based on their level of autonomous control. Following is a breakdown of the classification:

Level 0: Remotely Piloted Vehicle (RPV)
: At this level, UAVs are controlled entirely by human operators. The operators remotely pilot the drone and make all flight and mission decisions in real time.

Level 1: Execute Pre-planned Mission
: UAVs at this level are capable of executing preprogrammed missions autonomously. The mission parameters, waypoints, and flight paths are predefined by human operators, and the drone carries out the mission based on the given instructions.

Level 2: Changeable Mission

UAVs with level 2 autonomy possess the ability to adapt and modify their mission parameters while in flight. These drones can receive updated instructions from human operators or ground control stations and adjust their mission accordingly.

Level 3: Robust Response to Real-Time Faults/Events

At this level, UAVs have the capability to autonomously respond to real-time faults or unexpected events during their mission. They can identify and react to system failures, environmental changes, or other unforeseen circumstances, ensuring the continuation of the mission or safe return to base.

Level 4: Fault/Event Adaptive Vehicle

UAVs with level 4 autonomy exhibit a higher degree of adaptability. They can autonomously detect faults or events, diagnose the causes, and implement appropriate responses or workarounds to ensure mission completion or safe operation.

Level 5: Real-Time Multi-Vehicle Coordination

UAVs at level 5 possess the ability to coordinate and communicate with multiple other UAVs in real time. They can exchange information, share mission objectives, and collaborate on tasks to achieve coordinated operations.

Level 6: Real-Time Multi-Vehicle Cooperation

UAVs with level 6 autonomy go beyond coordination and actively cooperate with other UAVs to accomplish complex missions. They can dynamically allocate tasks, share resources, and adapt their behaviour based on the actions and needs of other cooperative vehicles.

Level 7: Battlespace Knowledge

At this level, UAVs possess a high level of situational awareness and understanding of the battlespace. They can collect and process information from various sensors, analyse the data, and provide actionable intelligence to human operators or other platforms.

Level 8: Battlespace Cognizance

UAVs with level 8 autonomy have advanced cognitive capabilities. They can interpret and comprehend complex situations, make decisions based on a deep understanding of the battlespace, and effectively carry out missions in dynamic and unpredictable environments.

Level 9: Battlespace Swarm Cognizance

UAVs at level 9 exhibit swarm intelligence and cognition. They can operate as part of a large networked swarm, communicate and cooperate with other swarm members, and collectively achieve complex missions with a high degree of autonomy.

9. List the criteria used by US DoD for classification of UAVs

Answer: The US Department of Defense (DoD) classifies Unmanned Aerial Systems (UAS) into different groups based on their Maximum Gross Take-Off Weight (MGTOW), normal operating altitude, speed, and representative UAS. Following is a description of each group:

1. Group 1:
   - MGTOW: 0–10 kg
   - Normal Operating Altitude: < 366m above ground level (AGL)
   - Speed: 185 km/hr
   - Representative UAS: Raven (RQ-11), WASP
2. Group 2:
   - MGTOW: 10–25 kg
   - Normal Operating Altitude: < 1067m AGL
   - Speed: 463 km/hr
   - Representative UAS: ScanEagle
3. Group 3:
   - MGTOW: < 600 kg
   - Normal Operating Altitude: < FL 180 (18,000 ft or 5,486 m)
   - Speed: 463 km/hr
   - Representative UAS: Shadow (RQ-7B), Tier II/STUAS
4. Group 4:
   - MGTOW: > 600 kg
   - Normal Operating Altitude: < FL 180 (18,000 ft or 5,486 m)
   - Any Speed
   - Representative UAS: Fire Scout (MQ-8B, RQ-8B), Predator (MQ-1A/B), Sky Warrior ERMP (MQ-1C)
5. Group 5:
   - MGTOW: > 600 kg
   - Normal Operating Altitude: > FL 180 (18,000 ft or 5,486 m)
   - Any Speed
   - Representative UAS: Reaper (MQ-9A), Global Hawk (RQ-4), BAMS (RQ-4N)

10. Classify UASs based on their applications

Answer:

a. Civil Applications:
   Photography: UAVs equipped with cameras are used for aerial photography and videography, providing unique perspectives and capturing images from above.

Construction: UAVs are utilized in the construction industry for surveying, mapping, and monitoring construction sites, allowing for accurate measurements and progress tracking.

Mining: UAVs can assist in mining operations by surveying and mapping mining sites, monitoring stockpiles, and conducting inspections of infrastructure and equipment.

Delivery and Logistics: UAVs are being explored as a means of delivering goods and providing logistical support, offering the potential for faster and more efficient delivery in certain areas.

Agriculture: UAVs equipped with specialized sensors and cameras are used for crop monitoring, precision agriculture, and for spraying pesticides or fertilizers on farmland.

Disaster Management: UAVs play a crucial role in disaster management, providing aerial views of affected areas, assessing damage, and aiding in search and rescue operations.

Surveillance: UAVs are used for surveillance purposes in various industries, including monitoring critical infrastructure, conducting border surveillance, and doing event security.

b. Environment Applications:

Air Quality Monitoring: UAVs equipped with air quality sensors can collect real-time data on air pollutants, helping to monitor and assess air quality in different regions.

Soil Monitoring: UAVs equipped with sensors can provide valuable data on soil conditions, moisture levels, and nutrient content, aiding in precision agriculture and land management.

Crop Monitoring: UAVs can capture high-resolution images of crops and analyse plant health, growth patterns, and pest infestations, enabling farmers to make informed decisions.

Water Inspection: UAVs can be used for monitoring water bodies, such as rivers, lakes, and reservoirs, to assess water quality, identify pollution sources, and monitor aquatic ecosystems.

Underwater Inspection: Specialized UAVs equipped with underwater cameras or sensors can conduct inspections of underwater structures, such as pipelines, dams, and marine infrastructure.

Mountain: UAVs are used for aerial mapping and monitoring in mountainous regions, assisting in land surveying, environmental research, and avalanche monitoring.

c. Defence Applications:

Combat Aircraft: UAVs designed for military applications, such as armed drones, are used for combat missions, surveillance, and reconnaissance in military operations.

War Zone Medical Supply: UAVs can be deployed to deliver medical supplies and provide support in war zones or areas with limited access due to conflicts or natural disasters.

Spying: UAVs equipped with advanced surveillance capabilities are employed for intelligence gathering, target acquisition, and monitoring of enemy activities.

Surveillance at Border: UAVs are utilized for border surveillance to monitor and secure national borders, detect illegal activities, and assist in border control and security operations.

Bomb Dropping: Military drones can be equipped with bombs or explosives for precision strikes and offensive operations in combat scenarios.

Missile Launching: Advanced military UAVs can launch missiles or precision-guided munitions, providing long-range strike capabilities and enhancing military firepower.

REFERENCES

Aabid, A., Parveez, B., Parveen, N., Khan, S., Raheman, M.A., Zayan, M., Shabbir, O. (2022). Reviews on Design and Development of Unmanned Aerial Vehicle (Drone) for Different Applications. *J Mech Eng Res Dev* 45, 53–69.

Abd Rahman, Y.A., Taghi Hajibeigy, M., Mahdi Al-Obaidi, A.S. (2018). Kean How Cheah, Design and Fabrication of Small Vertical-Take-Off-Landing Unmanned Aerial Vehicle, *MATEC Web Conf.* 152, 02023. 10.1051/matecconf/20181520202

Ahmed, F., Mohanta, J.C., Keshari, A. et al. (2022). Recent Advances in Unmanned Aerial Vehicles: A Review. *Arab J Sci Eng* 47, 7963–7984. https://doi.org/10.1007/s13369-022-06738-0

Alulema, V., Valencia, E., Cando, E., Hidalgo, V., Rodriguez, D. (2021). Propulsion Sizing Correlations for Electrical and Fuel Powered Unmanned Aerial Vehicles. *Aerospace* 8(7), 171. https://doi.org/10.3390/aerospace8070171

Czyba, R. et al. (2018). Construction Prototyping, Flight Dynamics Modeling, and Aerodynamic Analysis of Hybrid VTOL Unmanned Aircraft. *J Adv Transport*, Article ID 7040531-2018

Giordan, D., Adams, M.S., Aicardi, I. et al. (2020). The Use of Unmanned Aerial Vehicles (UAVs) for Engineering Geology Applications. *Bull Eng Geol Environ* 79, 3437–3481. https://doi.org/10.1007/s10064-020-01766-2

Jalled, F., Voronkov, I. (2016). Object Detection using Image Processing. https://doi.org/10.48550/arXiv.1611.07791

Jyoti, R.S.B. (2020). Classification of Unmanned Aerial vehicles: A Mirror Review. International Conference on Intelligent Engineering and Management (ICIEM). 10.1109/ICIEM48762.2020.9160179

Martinez, M. UAS Detection, Classification, and Tracking in Urban Terrain. Department of Electrical and Computer Engineering, Duke University. 10.1109/RADAR.2019.8835545

Mcfadyen, A., Mejias, L. (2016). A Survey of Autonomous Vision-Based See and Avoid for Unmanned Aircraft Systems. Progress Aerospace Sci 80, 1–17. [Online]. Available: www.sciencedirect.com/science/article/pii/S0376042115300208

Ozdemir, U. et al. (2014). Design of a Commercial Hybrid VTOL UAV System. *J Intelligent Robot Sys* 74, 371–393.

Reid, W.A., Albayati, I.M. (2021). Design of an Unmanned Aircraft System for High-Altitude 1 kW Fuel Cell Power System. *AS* 4, 353–363. https://doi.org/10.1007/s42401-021-00101-1

Singhal, G., Barrod, B., Mathew, L. (2018). Unmanned Aerial Vehicle Classification, Applications and challenger: A Review. Preprints. https://doi.org/10.20944/preprints201811.0601.v1

Ukaegbu, U., Tartibu, L., Okwu, M. (2021). Unmanned Aerial Vehicles for the Future: Classification, Challenges, and Opportunities. 2021 International Conference on Artificial Intelligence, Big Data, Computing and Data Communication Systems (icABCD). 10.1109/icABCD51485.2021.9519367

Yasin, J.N., Mohamed, S.A.S., Haghbayan, M.-H., Heikkonen, J., Tenhunen, H., Plosila, J. (2020). Unmanned Aerial Vehicles (UAVs): Collision Avoidance Systems and Approaches. *IEEE Access* 8, 105139–105155. 10.1109/ACCESS.2020.3000064

Chapter 6

Counter-UAS systems

OVERVIEW

The term "Counter-Unmanned Autonomous System" (C-UAS) refers to systems that are used to detect, track, and disable unmanned aerial systems. This chapter focuses on the need for C-UASs (Brooks, 2019), the role of a complete C-UAS, different types of C-UASs, and their limitations (Jean-Paul, 2020).

6.1 NEED FOR A COUNTER-UAS

The expanding use of civilian and wartime drones as a weapon has given rise to Counter-Unmanned Autonomous Systems (C-UAS) (Arthur, 2019). The need for C-UAS is evident from events reported in the recent wars where drone attacks are wide spread and has played crucial role in the war. C-UASs are small, and are inexpensive compared to the use of advanced air and missile defence systems. Air defence systems are primarily intended to detect and neutralize manned aircraft systems of the enemy. As a result, these systems are optimized to track and identify manned aircraft and are equipped to take down large, fast-moving objects. However, they may not be as effective in safeguarding against small, slow-moving unmanned aerial systems flying at low altitudes.

According to A. H. Michel (2019), there are as many as 537 counter-drone systems in the global market. Even though there has been an increase in knowledge regarding C-UASs technology globally, there is a need to study and mitigate plans based on local external factors.

A complete C-UAS should be able to detect, track, and intercept an unmanned aircraft. Some of the major UAS detection, tracking, and intercepting methods are as follows:

6.2 DETECTING, TRACKING, AND IDENTIFICATION TECHNIQUES

Some of the prominent Drone detection, tracking and identification techniques are depicted in Figure 6.1 and their methods are discussed subsequently.

DOI: 10.1201/9781003432685-6

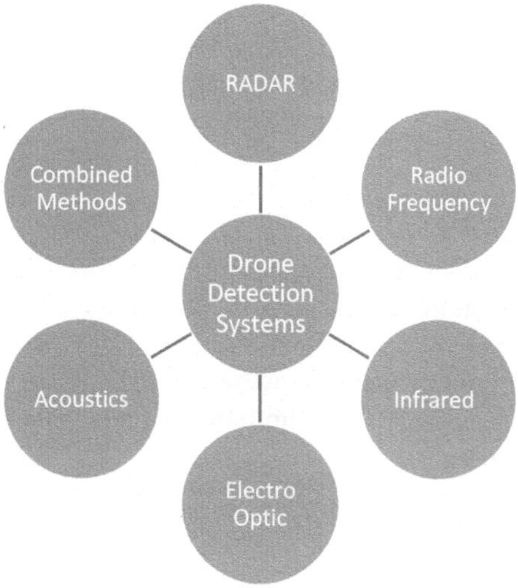

Figure 6.1 Different drone detection and tracking techniques.

6.2.1 Radar

Radar is used to detect the presence of small unmanned aircraft by their radar signature, which is generated when the aircraft encounters radio frequency pulses emitted by the detection element.

In the case of a radar system, artificial intelligence (AI)-based or image processing-based algorithms are used to distinguish between drones and other small, low-altitude flying objects, such as birds. Their detection and tracking capability depends on various factors such as environment and weather conditions, target type, and its size (Coluccia et al., 2021).

6.2.2 Radio Frequency (RF)

RF scanners and spectrum analysers identify the presence of drones by scanning for frequencies on which most drones are known to operate. Algorithms do pick out and geolocate RF-emitting devices in an area that are likely to be drones, but they are not very accurate. RF scanners can be effective at detecting the presence of a drone and probably identifying the type (Nemer et al., 2021).

6.2.3 Infrared (IR)

Thermal, or IR-based cameras are a good option to use in darkness. These cameras capture the heat signature of UAVs. While thermal cameras also have limitations around distinguishing drones from birds, however, heat signature databases make this distinction an easier challenge to resolve. Most of the commercial systems combine visual and thermal sensors to make them deployable in day, night, and twilight conditions (Hrúz et al., 2021).

6.2.4 Electro-Optical (EO)

Visual, or Electro-Optical (EO)-based cameras are used for the detection and tracking of UAVs in short ranges. A UAV can be detected and its path traced using a single camera by detecting motion cues, visual marks, and shape detectors. Multiple fixed ground cameras are also used in some anti-drone systems. All visual cameras have the natural limitation of being ineffective in darkness (Opromolla et al., 2019).

6.2.5 Acoustic

Acoustic sensors (Park, 2015) overcome an EO/IR sensor's challenge of line-of-sight requirements and the small size of UAVs. Typically, these systems capture the noise of UAV rotors and compare them with a library of sounds produced by known drones. The sound signatures of drones are highly affected by surrounding noise, wind direction, and ambient temperature.

6.2.6 RF jamming

RF jamming disrupts the radio frequency link between the drone and its operator by generating large volumes of RF interference. Once the RF link, which can include Wi-Fi links, is severed, a drone will usually either descend to the ground or initiate a "return to home" manoeuvre (Alhaj et al., 2019).

6.2.7 Combined sensors

Many systems integrate a variety of different sensor types to provide a more robust detection, tracking, and identification capability

6.3 INTERDICTION TECHNIQUES

In this section we discuss five popular drone interdiction techniques (Figure 6.2) viz. GNSS jamming, spoofing, dazzling, high Power Microwave, netting, collision drone and projectile.

6.3.1 GNSS jamming

Global Navigation Satellite Systems (GNSS) disrupt the drone's satellite link, such as the Global Positioning System (GPS) or the Global Navigation Satellite

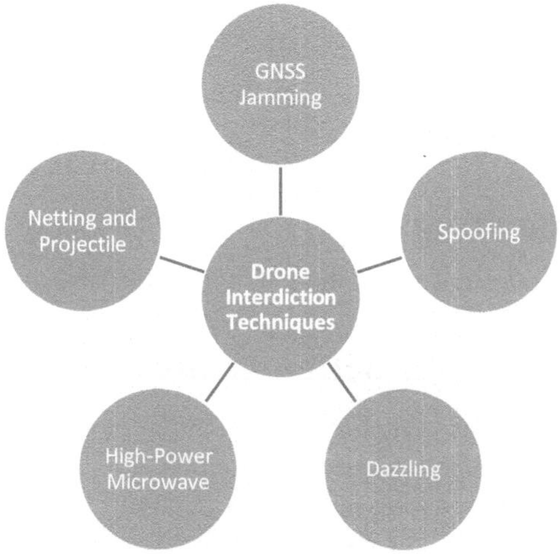

Figure 6.2 Drone interdiction techniques.

System (GLONASS), which is used for navigation. Drones that lose their satellite link will usually hover in place, land, or return to home (Maj Gen Harvijay Singh et al., 2020).

6.3.2 Spoofing

Spoofing allows one to take control of or misdirect the targeted drone by feeding it a spurious communications or navigation link, such as cyberattacks, protocol manipulation, and RF/GNSS Deception (Gudla et al., 2018).

6.3.3 Dazzling

Dazzling employs a high-intensity light beam or laser to "blind" the camera on a drone. The laser destroys vital segments of the drone's airframe using directed energy, causing it to crash to the ground (Lykou et al., 2020).

6.3.4 High-power microwave

A high-power microwave directs pulses of high-intensity microwave energy at the drone, disabling the aircraft's electronic systems (Chaari, 2021).

6.3.5 Nets

Nets are designed to entangle the targeted drone and/or its rotors.

6.3.6 Projectile

A projectile employs regular or custom-designed ammunition to destroy incoming unmanned aircraft.

6.3.7 Collision drone

A collision drone is designed to collide with the adversary drone.

6.3.8 Combined interdiction elements

A number of C-UASs also employ a combination of interdiction elements to increase the likelihood of a successful interdiction. For example, many jamming systems have both RF jamming and GNSS jamming capabilities in the same package. Other systems might employ an electronic system as a first line of defence and a kinetic system as a backup measure (Castrillo et al., 2022).

6.4 TECHNIQUES TO DESTROY DRONES OR RENDER THEM INEFFECTIVE

6.4.1 Fog bomb

A fog bomb is a kind of fogger to disable the cameras and other subsystems of rogue drones, especially intelligence-capturing drones. It is used for a few minutes at discrete intervals.

6.4.2 Water projector

High-pressure water projectors are employed to soak a rogue drone to disable and destroy it.

6.4.3 Prey birds

Large prey birds are trained to intercept rogue drones in mid-flight.

6.4.4 Slingshots

Slingshot shotguns can be used by trained security personnel to bring down rogue drones in visible ranges.

A C-UAS is required to have the capabilities of detecting the presence of a UAS, further locating and identifying its location, and then deploying measures to mitigate its effects. It requires interaction between distinct systems, and those systems and the ground operator. C-UASs make use of materiel systems;

tactics, techniques, procedures; and other approaches to prevent a UAS from accomplishing its mission. The kill chain of a C-UAS materiel system is as follows:

- → Detect, Locate, and Track the potential targets
- → Identify, Classify, and Evaluate targets as UASs
- → Identification of friend or foe
- → Engage and Neutralize UAS
- → Conduct Battle Damage Assessment
- → Clean-up and Recovery

6.5 CHALLENGES OF COUNTER-UASS

6.5.1 Detection effectiveness

The traditional detection systems might have technological challenges (Wang, 2021) when used in C-UAS operations:

- Radar systems may struggle to pick out small drones and UASs flying very close to the ground.
- Camera systems might confuse a drone with a bird or an airplane, and may also struggle in adverse weather with low visibility, or if the drone is backlit by a strong light source such as the sun.
- Electro-magnetic interference can degrade the detection capabilities of RF sensors. In urban environments, there are many potential sources of such interference, including communications antennae, two-way radios, telemetry systems, and even power lines and LED lights.
- Certain RF sensors, including some systems marketed as "passive," may likewise emit RF signals that could interfere with other communications, making them potentially dangerous to deploy in some environments.
- Radar, certain RF systems, and EO/IR sensors must have a direct line of sight with the intruding drone to make a detection. This could be particularly problematic in urban environments, where a drone may only appear within a sensor's line of sight for a couple of seconds before disappearing again.
- Some detection systems may only be capable of providing a rough estimate of an incoming drone's location.
- Certain flight patterns – most notably hovering and moving vertically – can make drones harder to detect with automated tracking algorithms applied to radar or camera data.
- Acoustic sensors rely on a library of sounds emitted by known drones, while RF detection systems only detect certain frequency bands in a pre-established library. Given the rapid rate at which drones are emerging on

the market and proliferating, even libraries that are updated often will never cover 100% of the drones that might be operating at any given time.

6.5.2 False negatives and false positives

A good C-UAS detection system is required to generate low levels of false negatives and false positives, which is difficult to achieve because C-UAS detection elements must be sensitive enough to detect all drones operating within the area of use, but systems that are too sensitive may create an overwhelming number of false positives, rendering the system unusable. Distinguishing true positives from false positives in a cluttered environment may require better-trained manpower.

6.5.3 Distinguishing legitimate and illegitimate drone use

In future operating environments where legitimate drone use is common, it will become increasingly important for C-UAS operators to be capable of differentiating between legitimate and rogue drones. Particularly, given the potential hazards of mitigating a drone in civilian environments, C-UAS operators will need to develop means to rapidly and reliably determine the threat level of an incoming UAV based on the limited information provided by existing detection technologies. A C-UAS that cannot tell the difference between allied and adversary unmanned aircraft could accidentally shoot down friendly drones.

6.5.4 Response window

C-UAS operators may have only a very brief window of time to decide whether an incoming drone is indeed malicious.

6.5.5 Interdiction hazards

C-UAS interdiction techniques can be dangerous in certain circumstances. Drones that have their flight interrupted by kinetic means may fall to the ground with considerable force. Even certain net-based systems that are equipped with a parachute to bring the ensnared drone down to the ground in a controlled manner may be risky if the parachute fails to deploy correctly, or if the interdiction occurs at a low altitude. Interdiction elements must be incredibly precise to hit a moving drone and could be dangerous to bystanders if they miss. Long-range effectors such as lasers and high-powered microwaves could pose a serious threat to aircraft operating above a targeted drone. Jamming systems, meanwhile, can interfere with legitimate communications links in their vicinity. The use of GPS jamming or spoofing systems, in particular, is especially dangerous in areas where other entities rely on reliable GPS navigation (for example, manned aircraft at an airport).

6.5.5.1 Interdiction effectiveness

Like detection systems, no interdiction system is 100% effective, as all interdiction systems have specific drawbacks. Some of these are the following: (Chiper et al., 2022)

- RF jamming systems do not affect drones that operate without an active RF link.
- Many signal jammers have a limited effective range of a few hundred metres, meaning that the system must be very close to the intruding drone to successfully mitigate it, and are not effective without a direct line of sight to the drone. Jammers that are capable of operating at long ranges and beyond line of sight must be significantly more powerful, but more powerful jammers also pose a higher risk of interference with legitimate communications.
- All kinetic systems may struggle against drones that are moving quickly or in unpredictable patterns. And when they do work as intended, they may destroy components of the drone that are necessary for forensic investigations.
- Spoofing systems, meanwhile, are technically very difficult to build and implement, and may not be universally effective against all drones. Unmanned aircraft that have been built with protected communication links could be resistant to spoofing attacks.

6.6 SOME OF THE COMMERCIALLY AVAILABLE SYSTEMS TO COUNTER ROGUE DRONES

6.6.1 RF/GNSS jamming systems

Jamming is the most common interdiction technique used. A few systems which are in use by different nations are discussed below. Also listed are the system specifications of Xpeller (Germany), RS500 (Israel), Ctrl+Sky (Poland), and AUDS (UK) respectively.

6.6.1.1 Ctrl+Sky

Ctrl+Sky has a scalable multisensor approach to ensure complete protection from rogue drones. The combination of radar, acoustic, vision, and RF sensors allows

Ctrl+Sky to minimize false alarms and detects even small drones at distances up to 3 km. The FMCW radar sensor operates in the X-band and uses MIMO technology to locate drones. The system can distinguish drones from other flying objects, such as birds. The acoustic sensor is based on an eight-element microphone array that uses a digital beam-forming technique in the acoustic domain to accurately sound sources in 3D space. An RF sensor detects a radio link between a drone and a remote control station by identifying RF signatures of Wi-Fi signals. The RF sensor detects a targeted radio signal, identifies it, and the software allows the neutralization of the rogue drone. By utilizing multiple RF sensors, the drone operator could also be localized. Ctrl+Sky jammer is used to neutralize drones by overpowering the drone's receivers (Svanström et al., 2022). The specification of Ctrl+Sky Drone detection is enlisted in Table 6.1.

Table 6.1 Specifications of Ctrl+Sky Drone Detection System

Ctrl+Sky Drone Detection & Neutralization System	
Detection Range	Up to 3,000 m (for micro-class drones sized DJI phantom IV)
Radar Beam Coverage	Sectional or omnidirectional
Radar Frequency	Default 10.5–10.6 GHx, optional 9.5–10.8 GHz
RF Power	Maximum 8W
RF Sensor	2.4 GHz, 5.8 GHz; other bands optional
Radar Modulation	FM, CW
Range Modulation Resolution	1–4 m (programmable)
Camera	PTZ, day and night vision, 30x optical zoom
Acoustic Sensor	8-microphone array
Jammer	• All currently functional global navigation satellite system (GNSS) GPS, GLONASS, Galileo, BeiDou in all operating bands:L1, L2, L5; 10 W jamming power • 2.4 GHz radio communication, 10 W jamming power • 5.8 GHz radio communication, 4 W jamming power • 433 MHz radio communication, 10 W jamming power • 868/915 MHz radio communication 10 W jamming power • any selected band between 400 MHz and 2800 MHz (e.g. 2G, 3G, 4G, LTE) 10 W jamming power
Platform	Ground-based stationary portable/Mobile
Power Consumption	Maximum 1,000 W
Power Supply	24 V sensors,230/110 V Central Processing Unit
Communication	Ethernet
Outdoor Weather Rating	IP 66 sensors, IP 55 central processing unit

6.6.1.2 *The Anti-UAV Defence System (AUDS)*

The Anti-UAV Defence System (AUDS) is a countermeasure against UAVs. It is designed to detect, track, classify, and disrupt threats from micro, mini, and large UAVs. The system consists of Blighter A400 Series Air Security Radar, Hawkeye DS, an EO Video Tracker, and a Directional RF Inhibitor.

The AUDS can be deployed in different terrain conditions and is capable of operating from fixed locations, as well as mobile platforms. It can detect micro UAVs up to a distance of 2 km, and mini or large UAVs at ranges up to 8 km. It has proved to be highly effective against swarm (Natalie, 2005) attacks and has successfully defeated approaching 2,000 drone sorties and has been tested against more than 60 types of drones including fixed-wing and quad-copters. Specification of AUDS is tabulated in Table 6.2.

Table 6.2 Specifications of anti-UAV defence system (AUDS)

Blighter A400 Series Air Security Radar	
Detection Range–	10 km
Minimum Target Sixe (RCS)	0.01 M^2
Frequency Band	Ku-band
Radar Type	E-scab Frequency Modulated Continuous Wave (FMCW) Doppler Surveillance Radar
Power Type	4 Watt
Azimuth Coverage	180^0 (standard)or 360^0 (optional)
Elevation Coverage	10^0(M10S antennas) or 20^0 (W20S antenna)
Elevation Adjustment	-40^0 to 30^0 using optional Blighter Radar tilting System (BRTS)
	Hawkeye DS and EO Video Tracker
Viper Dynamic Positioner	Azimuth: Continuous; Elevation -50^0 to + 60^0; Max speed 60^0 Per sec
Piranha 46HR Camera	Colour HD 2.3 MP; Auto Focus; Zoom: 30xopticial12x digital
Thermal Camera	Gen3cooled;Resolution:640x512pixel;wavelength: 3to5μm; Zoom: 240 to 1.80 FOV
EO Video Tracker	Visions 4ce digital video tracker and detector
Optical Disruptor	1.4^0 high-intensity beam

Directional RF Inhibitor

- High Gain quad-band antenna system (pentaband 5.8 GHz option)
- Disruption/inhibition delivers operational
- Custom inhibition waveforms specific to the threat
- Includes GNSS frequencies
- Software-defined intelligence RF inhibition
- Optimized disruption profiles

6.7 SPOOFING SYSTEMS

Out of 537 C-UASs, 31 systems have the spoofing capability (Arthur, 2019). Some of these are: AirShield (USA), ATENEA (Spain), drone-taker (Ireland), MESMER (USA), and Titan (USA).

6.7.1 MESMER

MESMER (n.d.) is a counter-drone platform that utilizes sophisticated detection and mitigation strategies which automatically identify, stop, redirect, land, or take total control of a target drone or other radio-controlled devices. MESMER's RF protocol manipulation takes advantage of weaknesses inherent in digital radio protocols. Its protocol manipulation is low power, operating below one watt. Low-power operation allows MESMER to affect only targeted radio-controlled devices – leaving non-targeted communication signals in the vicinity alone and avoiding signal fratricide. The technical specification of MESME is enlisted in Table 6.3.

6.7.2 Titan

Titan (Figure 6.3) is an autonomous, artificial intelligence-enabled counter-drone solution that forces drones to land safely without disrupting nearby communications or electronics. In less than five minutes, Titan can be set up with

Table 6.3 Technical specifications of MESMER counter-drone system

MESMER Counter Drone Solution	
Frequency	2.4–2.5 GHz, 5.725–5.875 GHz, 430–45 MHz, 902–928 MHz
Transmit Power	< 1 W (configurable)
Antenna	Array of 7 antennae
Range	4 km line of sight in 2.4 GHz band at < 1W transmit power, using omnidirectional antenna. Range varies depending on antenna type, transmit power, and terrain.
Operation Modes	Detection only, Auto-mitigation, Manual mitigation
Dimensions	19 Wx7Hx22Din
Volume	1.7 ft^2 (0.05 m^2)
Weight	60 lbs. (27.2 kg)
Operating Voltage	10–32VDC
Power Consumption	360W(12A at 30 VDC supply)
Temperature	50^0 F to 95^0 F (10^{0}C to 35^{0}C) (stand-alone)

Figure 6.3 Titan hemisphere of protection.

a hemisphere of protection beyond 3 km horizontally and 1,000 ft vertically. Engineered for fixed, mobile, and dismounted operation, Titan has been tested to meet MILSTD 461/810 requirements. Highly reliable detection with an extremely low false-positive rate, Titan successfully detects controller, video, and Wi-Fi links for individual drones and swarms.

Titan's escalating and configurable countermeasure approach can defeat drones without interfering with common communications (Wi-Fi, Bluetooth).

Regarding laser and dazzling systems, out of 537 C-UASs, 21 have laser and three have dazzling as interdiction capabilities (Arthur, 2019). In this section we will present three systems. The first is Lockheed Martin's HELIOS, which has both laser and dazzling interdiction capabilities. The second is LOCUST, another US C-UAS which uses laser as interdiction method. The third is Rafael's Drone Dome-L, which utilises laser as an interdiction technique along with RF and GNSS jamming.

6.7.3 HELIOS

The High Energy Laser with Integrated Opticaldazzler and Surveillance (HELIOS) system (n.d.) is a defensive weapon system designed to burn boats and shoot down unmanned drones. HELIOS is a 60-kilowatt laser system that can burn small speed boats and can torch unmanned aerial vehicles out of the sky. HELIOS can simply "dazzle" a UAV's electro-optical sensors, damaging them and preventing them from performing their mission. HELIOS has a long-range intelligence, surveillance, and reconnaissance capability. It is capable of high-definition, high-powered digital magnification, allowing a ship fitted with it to closely watch nearby threats.

LOCUST, a Low-Cost Counter-Unmanned Aerial System for Targeting (ATA, 2019), is a scalable, compact, rugged, and modular multi-mission system that has been designed to conduct effective and efficient detection, identification, management, and mitigation of Group-1 and Group-2 UAS threats. LOCUST was designed from the ground up to provide a capable and cost-effective defence against low-cost asymmetrical UAS threats. LOCUST detects and identifies UAS threats using active and passive RF and EO/IR sensor subsystems, and it negates the threats using intelligent electronic attack (EA) and high-energy laser (HEL) effectors. Each LOCUST delivers a layered detection and identification capability in a fixed emplacement configuration or on a mobile platform with HEL shoot-on-the-move capability. LOCUST is designed for installation on military or commercial platforms to defeat UAS and ISR (Intelligence, Surveillance, and Reconnaissance) threats, as well as provide ISR capability. It can be rapidly inserted into integrated battle management, command, and control (BMC2) architectures as well. The technical specification of LOCUST Counter Drone System is enlisted in Table 6.4.

6.7.4 Drone Dome-L

RAFAEL's Drone Dome-L C-UAS Laser System (Figure 6.4) (Robin, 2020) provides a modular and robust infrastructure, comprised of electronic jammers and sensors, allowing effective detection, full identification, and neutralization of multiple Micro and Mini UAV threats by employing its unique algorithms.

The baseline Drone Dome-L configuration system comprises a RADA RPS-42 S-band multi-mission hemispheric radar, a long-range surveillance system (SPEED ER), a communications package, and the C-Guard RD (rapid deployment) lightweight, fully programmable, portable jammer and NetSense Wideband detection sensor systems, all supported by advanced algorithms. The RPS-42 is a four-panel (non-rotating) tactical air surveillance system delivering 360° coverage in azimuth and 90° in elevation. Optimized to detect, track, and classify all classes of UASs, the RPS-42 Active Electronically Scanned Array (AESA) software-defined radar has a detection radius of 30 km – including the

Table 6.4 Technical specifications of LOCUST counter-drone system

LOUST Counter Drone Solution	
RF Direction Finding	• 0.1–10 km Range • 360⁰ Field of View (FOV) • "Silent" Detection
Radar Detection	• 0.1–5 km Range • 90⁰ Azimuth FOV • 60⁰ Elevation FOV • "Non-Silent" Detection
EO/IR Detection and Identification	• 5 km Detection • 3 km Identification • 1⁰–20⁰ FOV • "Silent" Detection
High Energy Laser (BEL) Effector	• > 1 kn Effectively • precision Aim Point • Target: Single Drone (> 5Targets/Minute)
Electronic Attack (EA) Effector	• > 3 Km Effectively • 600 Elevation FOV • 600 Azimuth FOV • Multi-Target simultaneous effectors • TargetC3Link+GPS (as allowed)
Power and Thermal (Auxiliary)	• Supports 10-hour duty day (continuous ISR, 1 hour Target Engagement) • Generator and shore power compatible for continuous operations

Figure 6.4 Drone Dome-L C-UAS Laser System.

detection of a minimum target size of 0.002 m^2 at a range of 3.2 km and at altitudes from 30 to 30,000 ft. The NetSense Wideband sensor detects a broad range of transmission types, including short bursts and wideband/narrowband signals, with continuous, rapid scanning in the 20 MHz to 6 GHz frequency range, while the C-Guard RD blocks Very High Frequency (VHF) and Ultra-High Frequency (UHF) UAV channels with Reactive Jamming (RJ) technology (with up to 400 W of total radio frequency output power), and targets all possible links on standard ISM (Industrial, Scientific, and Medical) and Wi-Fi bands. At this level of capability, the UAS threat is neutralized by activation of directional GPS/GNSS and radio frequency inhibitor/jammer devices.

SPEED ER is an advanced four axes gyro-stabilized extended-range surveillance and observation system, featuring a multiple sensor package comprising a cooled infrared thermal imaging camera (Medium-Wave Infrared – MWIR) with a continuous optical zoom lens, a Short-Wave Infrared (SWIR) camera with a continuous optical zoom lens, two Charge Coupled Device (CCD – day) cameras: one wide field-of-view, one extra narrow field of view, and integrated Video Motion Detection (VMD) and Automated Target Recognition (ATR) functions. The system includes a laser rangefinder and a laser pointer for target marking. Drone Dome-L leverages the scalable commercial off-the-shelf fibre-optic laser technology developed by RAFAEL. Mounted on a stabilized pan-and-tilt gimbal, the system features a high-energy laser beam director integrated with a mid-wave infrared EO sensor package – which includes coarse tracking, fine tracking, range measurement, aim point selection and update, and track file update and report functionalities – with a Field Of Regard (FOR) of 360° and elevation of 0°–60°. The Drone Dome-L system includes three operational phases: detection, identification, neutralization. In the detection phase, radar, RF and EO/IR sensors are used, either together or separately, providing both detection and accurate position of hostile drones. The system's EO sensor automatically initiates a drone classification and identification process, which provides the accurate position of the drone. One of the Drone Dome-L's unique capabilities is integrating laser technology for hard-kill capabilities. When the C4I performs a positive identification, the system allocates the target to the laser effector, which locks and tracks the target and performs hard-kill.

6.7.5 PHASER

Raytheon's PHASER (Donald, 1983) High-Power Microwave system (HPM) uses directed energy to down drones – single ones or swarms – at the speed of light. Operators focus a wide, arcing energy beam on drones that sends out a short, high-power burst of electromagnetic energy, destroying their electronics and dropping them simultaneously from the sky. The system uses microwaves to disable Group-1 and Group-2 drones, ones that are less than 55 pounds and fly at altitudes of 1,200 to 3,500 feet at speeds between 100 and 200 knots. PHASER is a high-powered microwave cannon that emits radio frequencies in a conical beam. It does not cook a drone with heat. Instead, the weapon disrupts or destroys the drone's circuits with a burst of overwhelming energy. It is not a thermal effect; it is an electric field effect that is basically imposed on the electronics either to upset or permanently damage them, and the effect is essentially instantaneous. Lasers need to fix on a target to burn it, but HPM pulses are shorter than one microsecond. Operators need positive identification before using the weapon, so the "kill chain" starts with detection from radar and then a camera or other sensor. The PHASER antenna uses that data to aim at the enemy drones

Net Systems: Out of 537 C-UASs, 27 employ nets to capture rogue drones Some of these systems are: SPARROWHAWK (Ireland), Sentinel (Switzerland), Anti-Drone Trap (Hong Kong), and Drone Catcher (Netherland).

6.7.6 SPARROWHAWK

The SPARROWHAWK is a C-UAS designed to capture, and recover intact, a rogue UAV and its payload safely (Figure 6.5). SPARROWHAWK will stop fully autonomous UAVs up to 20 kg, both rotary and fixed wing, and can be rearmed in just a few seconds for repeat sorties. Batteries can also be swapped

Figure 6.5 Raytheon's PHASER System.

in seconds, minimizing downtime. The SPARROWHAWK system is portable, reliable, quick to deploy, and easy to operate. It is comprised of a multi-copter COURIER UAV, weighted entanglement Net, parachute, safety interlocked deployment system, EOIR camera aiming, and a target selection module. SPARROWHAWK operates with the COURIER's enhanced video relay and remote monitoring systems so the pilot and observer/commander at the scene and remote personnel can monitor an incident even in poor light conditions. The technical specification of SPRROHAWK counter drone system of enlisted in Table 6.5

6.7.7 Sentinel Catch

Sentinel's Catch C-UAS intercepts and captures intruding drones using a net (Figure 6.6). It comes in two versions: Sentinel Catch which is launched from

Table 6.5 Technical specification of SPARROWHAWK

SPARROWHAWK C-UAV Module	
RESPONDER R900	
Frame Diameter	920 mm
Max Thrust (@Sea Level)	235 Newtons
Weight	6.5 kg (inc batteries, no payload
Duration	25 min (payload dependent)
Camera	EOIR Video Camera, Target Camera
SPARROWHAWK	
Max Capture Weight	20 kg
Net Capture Area	4–6 m^2
Lighting	9,000 Lumen (Optional)
TRANSIT CASE	
Dimensions (External)	673 mm (L)x673 mm (W)x 641 mm (H)
Weight (Empty)	18.6 kg
Material	Military-Grade Polythene
Colour	Black
GCS-Lite (ground Control Station)	
Dimensions	250 mm (L)x 400mm (W)x 200mm (H)
Control Frequency	2.4 GHz (other Frequencies Available
Radio Frequency (Video)	1.394 GHz (SD COFDM), 5.7 GHz (HD)
Video Recorder	In-built recorded on secure digital card (max32 GB)
Video Transmission Range	1,000 m depending on environment
Screen	Daylight Viewable
Weight	5 kg
Power Supply	Hot swappable dual battery@22.2 V
Battery Life Operation	Hot swappable batteries allow continuous operation

Figure 6.6 Skysc Sentinel Catch system (n.d.).

a mobile control vehicle and has a single propeller and 5km range, and Catch & Carry which is larger, has four propellers and which also captures unwanted drones in a net, but can stay in the air to carry what it has caught to a safe location. The Sentinel Catch uses a parachute to ground the intercepted drone. Both variants have a modular design allowing for upgrades to laser or tv-guided mode; multiple flight modes such as take-off, cruise and interception; missions can be aborted at any time. Applications include urban defence and airport defence.

Skysec's Sentinel Catch is actively powered by an electric motor and is capable of speeds up to 230 km/h and a range of up to 5 km. With an intelligent seeker head in the nose cone, only a rough 3D position of the target has to be provided to the sentinel. This can be achieved by a radar, visual-spotter, or any other drone detection system which fulfils the minimum precision requirement.

The intelligent flight controller (Santoso et al., 2018) can not only intercept moving targets but also fly complex customized flight plans like take-off, cruise, terrain follow, object avoiding, fly to safe zone, and more. This gives the operator the tools to secure even complex terrain with maximum protection and security.

6.7.8 Sentinel Catch & Carry

Skysec's Sentinel Catch & Carry (n.d.) is actively powered by four electric motors and is capable of speeds up to 126 km/h with a range of up to 2 km (Figure 6.7). It is equipped with the same seeker head as the Sentinel Catch and has the same interception workflow. In contrast to the Sentinel Catch, the Sentinel Catch & Carry, as the name suggests, not only catches the target but is also able to carry it to a predefined drop zone. The Technical Specification of Sentinal Catch System is enlisted in Table 6.6.

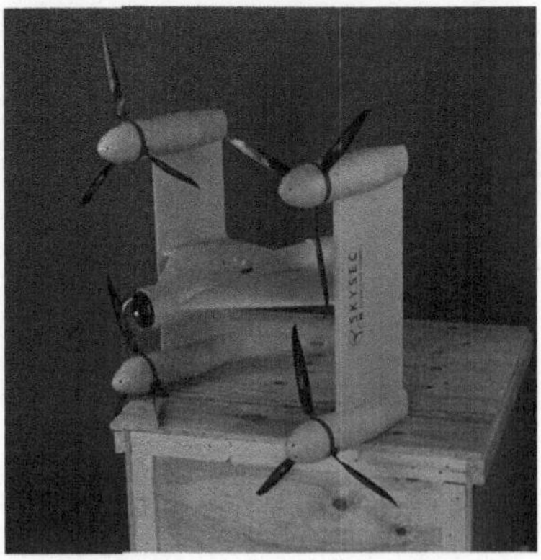

Figure 6.7 Skysc Sentinel Catch and Carry System (n.d.).

Table 6.6 Technical specifications of Sentinel Catch System

	Sentinel Catch	*Sentinel Catch & Carry*
Range	5 km	2 km
Speed	65m/s	65 m/s
Propulsion	Electric	Electric
Payload	Net & parachute	Net & hook 6 kg
Length	700 mm	700 mm
Weight	1.8 kg	8 kg
Features	• Modular design allows upgrade to laser-guided or television-guided mode • Multiple flight modes like take-off cruise and interception for maximum planning control and awareness. • Missions can be aborted anytime at either the current position or at a defined safe zone. • Parachute brings intercepted target and effectors safely to the ground to get minimal collateral damage	• Modular design allows upgrade to laser-guided or TV-guided mode • Multiple flight modes like take-off cruise and interception for maximum planning control and awareness • Missions can be aborted anytime at either the current position or at a defined safe zone. • Target can be caught and carried away to a defined drop zone to minimize possible collateral damage below interception

6.7.9 OVERWATCH

The OVERWATCH Interceptor-UX System detects a malicious or wayward drone, tracks it, and brings it to the ground using a safe and effective countermeasure. The system works by dispersing a cloud of latex-like countermeasures into the path of offending drones, stopping them in their tracks. The OVERWATCH Interceptor-UX includes four main components: the Interceptor-UX missile that creates the countermeasure cloud, uFDR detectors used to detect the UAV threats, the remote monitoring terminal which is used to identify the drone threat, and the base station used to launch the missile. The OVERWATCH system is safe to operate in urban environments, enables greater range, has no collateral damage, is effective against evasive manoeuvres and swarms, provides simplified logistics, and is lower cost than conventional kinetic or electronic warfare approaches. The OVERWATCH system detects, tracks, and mitigates an incoming threat within seconds using a rapid kinetic response. The world's lightest guided missile, weighing ≈700 grams and constructed from lightweight materials, does not cause collateral damage even if misused. The system stops wayward or malicious drones up to 2 km away and brings them decisively to the ground by deploying a cloud of environmentally friendly elastic countermeasures which become ensnared in the drone's propellers. For urban and airport applications, a custom parachute ensures the captured drone descends safely. The OVERWATCH Interceptor-UX missile can be mounted on a vehicle or vessel, but it can be also used by an individual soldier. It can offer a protection sphere of 1–4 km range. The system can be integrated into existing Battlefield Management Systems or used stand-alone. The technical specification of over watch Interceptor is enlisted in Table 6.7.

6.7.10 PITBULL AD

PITBULL AD (Figure 6.8) is an Ultra-Light Remote Weapon Station (ULRWS) with anti-drone capabilities that provides an ideal solution for detecting and tracking all types of drones during flight, enabling their neutralization and downing. PITBULL AD, which weighs only 70 kg, is easily installed on all manned and unmanned ground, air, and maritime platforms including ultra-light vehicles such as ATVs, as well as on fixed stations.

The system, which can integrate with any existing sensors, detects and tracks all types of drones in flight, can neutralize them via RF Intercept, and if needed, bring them down by either soft or hard kill. Following are the major features of the PITBULL AD ULRWS:

- → Drone Soft Kill Option – for downing a drone by jamming
- → Drone Hard Kill Option – for shooting down a drone
- → Ultra-Light (70 kg) supports 7.62 or 5.56 machine guns
- → Comprehensive Situational Awareness – day and night with "Point & Shoot" on screen interface

Table 6.7 Technical specifications of OVERWATCH Interceptor

OVERWATCH Interceptor-UX	
Guidance System	PN with OVERWATCH Triangulation Link Feed
Navigation System	GPS+ Inertial, Roll- Stabilized, Forward Canard control
Length	89.5 cm (Block (OA); 107.5 (Block OB)
Diameter	5.08 cm
Weight	452 gm (block OA); 710 gm (Block 0B)
Range	1.1 km (Block 0A);1.6 km (Block 0B)
Warhead	Proprietary Non-Lethal Anti-UAV arrestor countermeasure cloud
Trigger	Proximity Fuse
Maximum Speed	Block OA:165m/sec(Mach0.48);Block 0B:265m/sec (mach 0.79)
Motor	Solid propellant (160Ns. Block 0A;320N.s 2-stage-Block 0B)
Mounts	IGL (Ground launch), INLS (Naval Variant), IALS (UAV air-to-Air Mount
Field Upgradable	Yes • Block 1A/1B (RF seeker Guidance) • Block2- Extended Range Booster • Block 3-CIWSEMP (Close in Weapon System-EMP Warhead)

Figure 6.8 PITBULL AD remote weapon station.

6.8 COLLISION SYSTEMS

There are eight systems categorized as sacrificial collision drones by Michel (2019). Some of them are Coyote (USA), Howler (USA), Goshawk (Israel), and Drone Bullet (Canada, USA)

6.9 DRONE BULLET

Drone Bullet (Figure 6.9 (a) (b)) is a type of drone that's one-part quad-copter and one-part missile. Its goal is to knock other drones out of the sky. The Drone Bullet leverages machine vision and a 4-km range to seek and destroy other unmanned aircraft. It weighs 910 grams, and the array of propellers gives it ample manoeuvrability. It can be launched by hand, and the operator only has to indicate a target. From there, the on-board software and cameras track the other drone and aim for a collision.

The Drone Bullet can even identify different types of unmanned aircraft and adjust its tactics accordingly. It does not carry an explosive charge, but rather smacks into the offending drone at high speed. Even a glancing blow is enough to neutralize the target when impacted at up to 350 km/hr. The Drone Bullet is built to be durable, so it might survive destroying its target. In that case, the drone has the ability to recalibrate and pursue a second objective. AerialX demoed the Drone Bullet for the US Special Operations Command (USSOC). During the test at Fort Bragg, the Drone Bullet tracked and destroyed various drones from small to large. Figure 6.9: Drone bullet (a) on the Fly (b) Deployed on Ground.

Figure 6.9 Drone Bullet.

Figure 6.10 Birds of prey.

6.10 BIRDS OF PREY

In an entirely different approach to C-UASs, the Dutch firm Guard from Above trains large birds of prey to intercept rogue drones in mid-flight (Figure 10.10). According to Guard from Above, eagles – which wear protective shin guards in order to shield their legs from the drone's rotors – have a 95% intercept rate, which is likely higher than many mechanical kinetic alternatives

6.11 ADVANCED COUNTER-UAS TECHNOLOGIES

6.11.1 Holographic radar

The Gamekeeper 16U Holographic C-UAS radar (Figure 6.11) is able to detect, track, and classify small UAS in 3D to a range of 5 km. Unlike the conventional scanning radars, it has no moving parts and instead continuously "floodlights" the entire field of view. Using an array of fixed receivers, a 3D image of the environment is formed. This allows the 3D location and velocity of every detected object to be provided. Powerful algorithms distinguish targets of interest from background clutter and classify them as UASs or other targets. As a software defined radar, Gamekeeper can quickly be adapted to deal with new environments and target types, based upon a modular and scalable hardware platform. The technical Specification of Holographic Radar System is depicted in Table 6.8.

Figure 6.11 Holographic CUAS radar system.

Table 6.8 Technical specifications of holographic radar system

Gamekeeper 16U C-UAS Holographic Radar	
Range	5 km
Azimuth	90⁰
Elevation	30⁰
Max altitude	900 m
Target size	0.01 m^2 RCS @ 5 km range
Classification	UAS, Aircraft, Ground Vehicle, Surface Vessel, Bird
Frequency	L- Band, 1.255 GHz
Power	2kW

6.12 SUMMARY

As discussed earlier, in effective C-UAS technologies, many types of sensors are required to be configured for detection, and multiple interdiction techniques are to be integrated for effective neutralization. For a comprehensive system design, the following factors are to be integrated into the design:

- Multisensor data fusion in real time, Image processing, Implementation of artificial intelligence (AI) algorithms for classification of objects and detection of RF frequency, and so forth.
- Integrated secured Command, Control, Communications, Computers, and Intelligence (C4I) with shooter or jamming system requires Integrated IFF operations capability for a drone war situation.
- Weather conditions and environmental factors
- Modular-based compatible subsystem and system
- Minimum weight and size (min form factor)

After detection and tracking of a UAV, in military scenarios, it may be required to identify it as a friend or foe. The interdiction of an adversary's UAV, if it is built on intelligent decision-based system, might use one of the integrated techniques. These are broadly classified as:

- Protocol-based interdiction
- Sensor-based interdiction
- Interdiction using jammers
- Physical interdiction

In addition to the detection and interdiction, there are certain recommendations for incorporation in SOPs (Standard Operating Procedures) in case of an unexpected UAV spotting. Actions to be incorporated into the operational documents are:

- Find pilots – Most drones are remotely piloted and given the range of RF operation and endurance, the pilot can be found in the vicinity.
- Find dropped objects – Some drones might drop small explosives or harmful payloads for disrupting activities.
- Close window shutters – To circumvent the intelligence intent of drones. Lead people to safety.
- Lock doors and gates – Drones once interdicted with soft technologies like jamming can exhibit unpredictable behaviour. These include: the drone makes a controlled landing in its current position.

QUESTIONS AND ANSWERS

1. What is a counter-UAS system? What is the necessity of a counter-UAS system?

Answer: A counter-UAS system refers to systems that are used to detect, track, and disable unmanned aerial systems. These are systems capable of lawfully and safely disabling, disrupting, or seizing control of an unmanned aircraft or unmanned aircraft system.

Counter-UASs are becoming increasingly necessary due to the growing use of drones for a variety of purposes, including military operations, surveillance, and civilian applications such as package delivery and aerial photography, which makes drones easily usable for carrying out malicious acts such as terrorist attacks, conducting espionage, or invading privacy. Without effective counter-UAV measures, drones can pose a significant threat to security and safety.

2. What are different techniques of detecting, tracking, and identifying a UAV?

Answer: Different techniques used for detecting, tracking, and identifying a UAV are as follows:

- Radar: Radar is one of the most widely used techniques for detecting UAVs. It works by sending out electromagnetic waves that bounce off objects and return to the radar. The radar can then detect the presence of a UAV based on the return signal.
- Acoustic sensors: Acoustic sensors use sound waves to detect the presence of UAVs. They work by picking up the sound of the UAV's propellers or motor and can be effective in detecting low-flying UAVs.
- Optical sensors: Optical sensors, such as cameras and telescopes, can be used to track and identify UAVs. They work by detecting the visible light or infrared radiation emitted by the UAV.
- Radio Frequency (RF) sensors: RF sensors can be used to detect the communication signals between a UAV and its operator. By analysing the RF signals, it is possible to determine the location of the UAV and track its movements.
- Drone-mounted sensors: Some UAVs are equipped with sensors that can be used to detect and track other UAVs. These sensors can include cameras, radar, and RF sensors.
- Artificial Intelligence (AI): AI can be used to analyse data from multiple sensors to detect and track UAVs. AI can be trained to recognize patterns and anomalies in sensor data, making it possible to identify UAVs even in complex and crowded environments.

3. List some common drone detection techniques.

Answer: Same as above

4. What are different challenges of counter-UASs?

Answer:

- Detection range: Drones can be difficult to detect at long ranges, especially if they are small or low-flying. This can make it difficult to detect and track drones before they enter a restricted area.
- Sensor limitations: Different sensors can have limitations, such as line-of-sight requirements, limited field of view, or difficulty detecting low-flying drones. This can make it challenging to get a complete picture of the drone's location and trajectory.
- Signal interference: Drone operators can use countermeasures, such as jamming or spoofing, to disrupt C-UASs. This can make it difficult to track or neutralize a drone.
- False positives: C-UASs can sometimes mistake other objects or signals for a drone, leading to false alarms. This can waste time and resources and reduce the effectiveness of the system.
- Cost: C-UASs can be expensive to acquire, operate, and maintain. This can be a significant challenge, especially for small businesses or organizations that do not have the budget to invest in expensive technology.
- Legal and ethical considerations: Using C-UASs can raise legal and ethical considerations, such as privacy concerns and compliance with local laws and regulations. It is essential to ensure that the use of C-UASs is within the legal and ethical boundaries.

5. What are false negatives and false positives in the detection of a drone?

Answer: C-UASs can sometimes mistake other objects or signals for a drone, leading to false alarms. This can waste time and resources and reduce the effectiveness of the system.

A false negative is an error in which a test or evaluation fails to detect a UAV that is actually present. In other words, it occurs when a test or evaluation incorrectly gives a negative result for a condition that is present.

6. How to distinguish between legitimate and illegitimate drone use?

Answer:

- Intention: The intention behind the drone use is an important factor in determining whether it is legitimate or not. For example, using a drone for search and rescue, or for aerial photography is generally considered legitimate, while using a drone for smuggling or surveillance can be considered illegitimate.

- Location: The location where the drone is being flown can also be an important factor in determining whether it is legitimate or not. Flying a drone in restricted airspace or near sensitive facilities, such as airports or government buildings, can be considered illegitimate.
- Licensing and regulations: Many countries have regulations for drone use, including licensing requirements, operating restrictions, and privacy guidelines. If the drone use is in compliance with these regulations, it is likely to be considered legitimate.
- Behaviour of the operator: The behaviour of the drone operator can also provide clues about whether the drone use is legitimate or not. For example, if the operator is flying the drone in a reckless or dangerous manner, it is more likely to be considered illegitimate.
- Payload: The payload carried by the drone can also be a factor in determining whether it is legitimate or not. For example, carrying cameras or sensors for scientific or research purposes is more likely to be considered legitimate, while carrying weapons or hazardous materials is more likely to be considered illegitimate.

7. With respect to C-UASs, what is/are (a) Response Window, (b) Interdiction Hazards?

Answer: C-UAS operators may only have a very brief window of time to decide whether an incoming drone is indeed malicious. This is called a response window.

Interdiction hazards refer to the risks or dangers associated with using interdiction techniques to counter UASs. Interdiction techniques can include kinetic means, such as shooting the drone down, or non-kinetic means, such as using jammers or net-based systems.

8. Interdiction Effectiveness

Answer: Interdiction effectiveness refers to the ability of a C-UAS to successfully detect, track, and neutralize a target drone. The effectiveness of a system depends on various factors, including the type of system used, the capabilities of the system, and the environmental conditions in which the system is deployed.

The effectiveness of a system can be improved through continuous improvement and updating of the system's capabilities, as well as training for the operators of the system. Additionally, collaboration and information sharing among different agencies and organizations can help to enhance interdiction effectiveness by enabling a more coordinated and comprehensive approach to countering unauthorized drone activity.

9. List some of the commercially available C-UASs to counter rogue drones.

Answer:

- RF/GNSS Jamming Systems: Such as Xpeller (Germany), RS500 (Israel), Ctrl+Sky (Poland), and AUDS (UK).
- The Anti-UAV Defence System (AUDS): The system consists of Blighter A400 Series Air Security Radar, Hawkeye DS, an EO Video Tracker, and a Directional RF Inhibitor.
- Spoofing Systems: MESMER, Titan, HELIOS, LOCUST, Drone Dome-L, PHASER, SPARROWHAWK, Sentinel Catch, Sentinel Catch & Carry, Overwatch

10. What is the GNSS jamming technique? How can GNSS jamming be used as an anti-drone system?

Answer: Global Navigation Satellite System (GNSS) jamming is a technique used to disrupt GNSS signals. This is done by broadcasting a high-powered signal in the same frequency band as the GNSS signal, causing the receiver to be unable to process the signal accurately. GNSS jamming can be used to prevent drones from flying in restricted areas or to force them to land.

To use GNSS jamming as an anti-drone system, a jamming device is placed in the vicinity of the area to be protected. When a drone enters the area, the jammer broadcasts a signal that interferes with the GNSS signals being received by the drone's navigation system. This can cause the drone to lose its ability to navigate, forcing it to land or return to its starting point.

While GNSS jamming can be an effective way to counter drones, it is important to note that it can also have unintended consequences. Jamming signals can interfere with legitimate GPS navigation systems, such as those used by aircraft, ships, and emergency services. Additionally, it may be difficult to use GNSS jamming effectively in urban areas where buildings can block the signal and reflect it in unpredictable ways. Therefore, it is important to use GNSS jamming in a carefully controlled manner, with proper planning and coordination with relevant authorities.

11. What is GNSS spoofing? How can GNSS spoofing be used to stall the mission of a UAV?

Answer: GNSS spoofing is a technique used to deceive a GNSS receiver by broadcasting fake signals that mimic the authentic GNSS signals. The goal of

GNSS spoofing is to make the receiver think it is receiving legitimate GNSS signals, while in reality it is receiving false signals that can be used to manipulate the receiver's location information.

To use GNSS spoofing as a counter-drone system, a spoofer device is placed near the area to be protected. The spoofer device generates fake GNSS signals that can be used to fool the drone's navigation system into thinking that it is at a different location than it actually is. This can cause the drone to lose its ability to navigate and either crash or return to its starting point.

Spoofing can be particularly effective against drones that use GNSS as their primary means of navigation. However, spoofing can also have unintended consequences, such as interfering with legitimate GPS navigation systems and causing potential safety hazards.

It is important to use GNSS spoofing in a carefully controlled manner, with proper planning and coordination with relevant authorities. Additionally, it is important to note that modern drones may have additional sensors and navigation systems that can help them navigate in the event of GNSS spoofing. Therefore, it is necessary to take a multilayered approach to counter-drone systems that combines different techniques to improve overall effectiveness.

12. What is dazzling? How can dazzling can render a UAV dysfunctional?

Answer: Dazzling is a technique used to temporarily blind a drone's camera or pilot, rendering the UAV dysfunctional. Dazzling can be achieved using various sources of light, such as high-powered lasers or even simple flashlights. When a drone is flying, its camera is usually the primary source of navigation and situational awareness for the pilot or autonomous system. By shining a bright light, such as a laser, at the drone's camera, the camera's sensors can be overwhelmed and disrupted, causing the drone to lose its ability to navigate and maintain its position. Similarly, if a pilot is operating the drone and is hit with a dazzling light source, they may be unable to maintain control of the UAV, causing it to crash.

Dazzling can be an effective counter-drone technique, particularly against drones that are being piloted visually. However, it is important to note that some drones may have sensors and backup navigation systems that can help them maintain their position, even if the camera is temporarily disrupted. Additionally, dazzling can be dangerous if not used properly and may cause harm to innocent bystanders or nearby aircraft if the drone is not brought down safely.

13. What are combined interdiction techniques? List some of the popular interdiction techniques used in anti-drone activities.

Answer: Combined interdiction techniques refer to the use of multiple counter-drone technologies and methods to increase the overall effectiveness of anti-drone activities. By combining different techniques, such as detection, tracking, identification, and interdiction, it is possible to create a comprehensive anti-drone system that can effectively neutralize UAV threats.

Some popular interdiction techniques used in anti-drone activities include:

- Kinetic methods: Using physical means, such as firearms or nets, to intercept and bring down the UAV.
- Radio frequency (RF) jamming: Jamming the drone's communication channels with the operator or GPS signals, causing it to lose control or fly erratically.
- GNSS spoofing: Spoofing the drone's GPS signal to trick it into thinking it is in a different location or altitude, causing it to deviate from its course or even crash.
- Directed energy weapons: Using high-powered lasers or microwaves to disrupt the drone's electronics or damage its components, causing it to malfunction or crash.
- Dazzling: Using bright lights or lasers to blind the drone's camera or pilot, rendering it dysfunctional.
- Drone capture: Using capture devices, such as nets or cages, to physically capture the drone and prevent it from flying.

14. List four different techniques to destroy drones and render them ineffective.

Answer:

- **Fog Bomb:** This is a kind of fogger to disable the cameras and other subsystems of rogue drones, especially intelligence-capturing drones. It is used for a few minutes at discrete intervals.
- **Water Projector:** High-pressure water projectors are employed to soak a rogue drone to disable and destroy it.
- **Prey Birds:** Large prey birds are trained to intercept rogue drones in mid-flight.
- **Slingshots:** Slingshot shotguns can be used by trained security personnel to bring down rogue drones in visible ranges.

15. What are some challenges in a counter-UAV system?

Answer:

- Detection Effectiveness: Counter-UAV systems may face challenges in detecting drones, particularly small drones, due to their low radar cross section and advanced flight capabilities.
- False Negatives and False Positives: Counter UAV systems may generate false negatives, failing to detect a drone, or false positives, detecting

an object that is not a drone, which can lead to unnecessary disruption or alarm.

- Distinguishing Legitimate and Illegitimate Drone Use: It can be difficult to differentiate between drones being used for legitimate purposes, such as aerial photography or surveying, and drones being used for nefarious activities.
- Response Window: Counter-UAV systems must be able to respond quickly to a detected drone to effectively interdict it, which can be challenging due to the speed and agility of drones.
- Interdiction Hazards: Certain interdiction techniques, such as kinetic means, can pose safety risks to bystanders or damage infrastructure, while long-range effectors and jamming systems can interfere with legitimate communications and navigation systems.

16. What is UAV interdiction? How it is used to counter a UAV?

Answer: UAV interdiction refers to the act of disrupting or disabling a UAV in order to prevent it from completing its mission. Interdiction techniques can include jamming, spoofing, kinetic means, net-based systems, and dazzling. The choice of technique depends on the UAV's size, altitude, and flight capabilities, as well as the surrounding environment and potential hazards. Interdiction systems must be precise, reliable, and effective in order to safely and efficiently counter UAV threats. Successful UAV interdiction can help protect critical infrastructure, airspace, and public safety.

17. Tabulate the generic specifications of a counter-drone system.

Answer:

Detection Range	10 km
Minimum Target Sixe (RCS)	0.01 M^2
Frequency Band	Ku-band
Radar Type	E-scab Frequency Modulated Continuous Wave (FMCW) Doppler Surveillance Radar
Power Type	4 Watt
Azimuth Coverage	180^0 (standard) or 360^0 (optional)
Elevation coverage	10^0 (M10S antennas) or 20^0 (W20S antenna)
Elevation Adjustment	-40^0 to 30^0 using optional Blighter Radar tilting System (BRTS)
	Hawkeye DS and EO Video Tracker
Viper Dynamic Positioner	Azimuth: Continuous; Elevation -50^0 to $+60^0$; Max speed 60^0 Per sec

Piranha 46HR Camera	Colour HD 2.3 MP; Auto Focus; Zoom: 30xopticial12x digital
Thermal Camera	Gen3cooled; Resolution:640x512pixel;wavelie ngth:3to5µm; Zoom: 240 to 1.80 FOV
EO Video Tracker	Visions 4ce digital video tracker and detector
Optical Disruptor	1.4^0 high-intensity beam

REFERENCES

Adnan A.H. 1 1 TUSUR, Tomsk, Russia. 2019; 1–2: 241–244.

Brooks S., Kouhestani C. Counter-Unmanned Aircraft Systems Market Survey. Department of Homeland Security, Prepared by Sandia National Laboratories; 2019 Feb.

Castrillo V.U, Manco A., Pascarella D., Gigante G. A Review of Counter-UAS Technologies for Cooperative Defensive Teams of Drones. Drones. 2022; 6(3):65. https://doi.org/10.3390/drones6030065

Chaari M.Z. High Power Microwave for Knocking Out Programmable Suicide Drones. Security and Defence Quarterly. 2021;34(2):68–84. doi:10.35467/sdq/135068

Chiper F.-L., Martian A., Vladeanu C., Marghescu I., Craciunescu R., Fratu O. Drone Detection and Defense Systems: Survey and a Software-Defined Radio-Based Solution. Sensors. 2022; 22(4):1453. https://doi.org/10.3390/s22041453

Coluccia A., Fascista A., Schumann A., Sommer L., Dimou A., Zarpalas D., Méndez M., de la Iglesia D., González I., Mercier J.P., Gagné G., Mitra A., Rajashekar S. Drone vs. Bird Detection: Deep Learning Algorithms and Results from a Grand Challenge. Sensors (Basel). 2021 Apr 16;21(8):2824. doi:10.3390/s21082824. PMID: 33923829; PMCID: PMC8072977.

Donald J. Sullivan; "high power microwave generation from a virtual cathode oscillator (vircator)", *IEEE Transactions on Nuclear Science*, Vol. NS-30, No. 4, August 1983. Mission Research Corporation; Albuquerque, New Mexico 87106.

Franta N.R. Swarm Intelligence for Autonomous UAV Control. Naval Post Graduate School, Monterey, California; 2005 June.

Gudla C., Rana M., Sung A. Defense Techniques Against Cyber Attacks on Unmanned Aerial Vehicles., International Conference on Embedded Systems, Cyber-physical Systems, & Applications, 2018.

Hrúz M., Bugaj M., Novák A., Kandera B., Badánik B. The Use of UAV with Infrared Camera and RFID for Airframe Condition Monitoring. Applied Sciences. 2021; 11(9):3737. https://doi.org/10.3390/app11093737

Lykou G., Moustakas D., Gritzalis D. Defending Airports from UAS: A Survey on Cyber-Attacks and Counter-Drone Sensing Technologies. Sensors. 2020; 20(12):3537. https://doi.org/10.3390/s20123537

Maj Gen Harvijay Singh S.M. Countering Threats from Drones – 2 The Military Domain: Measures against Military/Rogue Drones. South Asia Defence & Strategic Review. 2020 May–Jun; 14(2):50.

Michel A.H. Counter-Drone Systems, 2nd ed.; 2019 Dec. Center for the Study of the Drone at Bard College, https://dronecenter.bard.edu/projects/couter-drone-systems-project/counter-drone-systems-2ndedition/.

Nemer I., Sheltami T., Ahmad I., Yasar A.U.-H., Abdeen M.A.R. RF-Based UAV Detection and Identification Using Hierarchical Learning Approach. Sensors. 2021; 21(6):1947. https://doi.org/10.3390/s21061947

Opromolla R., Inchingolo G., Fasano G. Airborne Visual Detection and Tracking of Cooperative UAVs Exploiting Deep Learning. Sensors. 2019; 19(19):4332. https://doi.org/10.3390/s19194332

Park S., Shin S., Kim Y., Matson E.T., Lee K., Kolodzy P.J., Slater J.C., Scherreik M., Sam M., Gallagher J.C., Kolodzy P., Fox, B., Hopmeier M. Combination of Radar and Audio Sensors for Identification of Rotor-Type Unmanned Aerial Vehicles (UAVS). Sensors. 2015; 1-4. 10.1109/ICSENS.2015.7370533

Santoso F., Garratt M.A., Anavatti S.G. State-of-the-Art Intelligent Flight Control Systems in Unmanned Aerial Vehicles. IEEE Transactions on Automation Science and Engineering. 2018 April; 15(2), 613–627. doi:10.1109/TASE.2017.2651109

SKYSEC Counter Drone Systems. n.d. www.skysec.ch/.

Svanström F., Alonso-Fernandez F., Englund C. Drone Detection and Tracking in Real-Time by Fusion of Different Sensing Modalities. Drones. 2022; 6(11):317. https://doi.org/10.3390/drones6110317

Wang J., Liu Y., Song H. Counter-Unmanned Aircraft System(s) (C-UAS): State of the Art, Challenges, and Future Trends. IEEE Aerospace and Electronic Systems Magazine. 2021; 36:4–29. 10.1109/MAES.2020.3015537

Yaacoub J.-P. et al. Security Analysis of Drones Systems: Attacks, Limitations, and Recommendations. Elsevier 2020 May 8.

Chapter 7

Design criteria of a UAV for photogrammetry and ISTAR applications

OVERVIEW

An Unmanned Aerial Vehicle (UAV) is a multi-role avionic system used for many applications. In this chapter we discuss various aspects of UAV as a platform for Photogrammetric and ISTAR application. Photogrammetry is the art and science of locating and measuring the geometric properties of spatial objects from photographs captured remotely using sensors from air and space.

This chapter discusses the design of a new integrated system combining the capability of three different technologies, namely, UAV, photogrammetry, and Artificial Intelligence (AI) to carry out Intelligence, Surveillance, Target Acquisition, and Reconnaissance (ISTAR) applications (Panigrahi, 2020). The design criteria of such a UAV are proposed in the form of min-max conditions of these associated technologies. This chapter also discusses the improved photogrammetric applications that can be performed using a UAV and AI. The Size, Weight, Area, and Power (SWAP) constraints in designing such a UAV for photogrammetric applications are discussed in this chapter (Colomina, 2014).

7.1 INTRODUCTION

The Unmanned Aerial System (UAS) is a generic class of aerial vehicles managed and navigated without a human pilot. With the advanced research in avionics, remote sensing, and sensor technology, the multi-role capability of this class of systems has evolved rapidly, resulting in numerous applications. This has brought about in the development of different types of UASs. A review of UAS development (Brain 2015; Geng 2014) reveals vehicles such as unmanned aerial vehicles with ground pilots, popularly known as Unmanned Aerial Vehicles (UAVs); drones (UAVs without a ground pilot); and nano, micro, and mini UAVs of varying endurance, mass, and range. Typically, UAVs employed in different applications are of mass 1–2 kg, 5–10 kg, and 10–150 kg respectively. Therefore, a UAV which is a generic multirole avionic system can be classified based on many different criteria such as (a) Application meant for, (b) Type of control and guidance system used, (c) Range and Altitude of airspace it

DOI: 10.1201/9781003432685-7

covers, (d) Type of take-off and landing process, (e) Fuelling and power system employed, (e) Weight of the craft and weight of payload it can carry, (f) Flight endurance of operation and so forth.

UAV (Brain 2015; Geng 2014) is an acronym for Unmanned Aerial Vehicle, which is an aerial platform with no pilot on board. UAVs can be remote-controlled aircraft (e.g. flown by a pilot at a ground control station), or can fly autonomously based on preprogrammed flight plans or more complex dynamic automation systems. Moreover, a drone is a fully automatic UAV with autonomous take-off, landing, and navigation in the air. Often, a drone can operate in collaboration with other drones taking navigation commands from other similar UAS.

Photogrammetry (Panigrahi & Sabitra 2018; Goshtaby 2005) as defined by the American Society for Photogrammetry and Remote Sensing is the art, science, and technology of obtaining reliable information about physical objects and the environment through processes of recording, measuring and interpreting photographic images and patterns of recorded radiant electromagnetic energy and other phenomena. Traditionally, photogrammetry analysis and measurements are carried out on images of the Earth's surface obtained by sensors deployed in the belly of an airplane or remote sensing satellites.

Generally, photogrammetry has been associated with Remote Sensing (RS) carried out using satellites or manned aircraft with a set of airborne sensors capturing images of the Earth's surface (Aasen et al., 2018). Ground computing is performed for the classification and estimation of Earth parameters using photogrammetric techniques. In the last decade, the increasing developments and improvements in unmanned platforms, together with the development of sensing technologies installed on board such as platforms, have provided excellent opportunities for different photogrammetric applications. Indeed, they can offer high versatility and flexibility, as compared to airborne systems or satellites, and can operate rapidly without planned scheduling. RS operations using UAVs can be carried out where the safety of life is at stake due to disastrous situations in the near ranges. UAVs can fly at low altitudes and slowly, with the ability to acquire high spatial and temporal resolution data (Mohsan et al., 2023). Of late, UAVs have emerged as the most sought-after platform for aerial photogrammetry over traditional aerial photographic platforms such as aeroplanes or satellites because of the following advantages.

7.2 ADVANTAGES OF UNMANNED AERIAL VEHICLES (UAVS) FOR PHOTOGRAMMETRY OPERATIONS

This section presents the advantages of UAV photogrammetry compared to traditional satellite remote sensing data. The terrain information generated by UAV photogrammetry methods provides a good representation of the topography. The usefulness of drone-based photogrammetry for terrain analysis is (a) the generation of dense 3D point cloud of the terrain surface, (b) digital surface and terrain model, (c) orthomosaic, (d) volume calculation, (e) contour line generation, (f) 3D textured model, (g) reflectance maps, (h) application maps, and (i) thermography – generate a radiometrically accurate map with a temperature value of each pixel.

7.2.1 Outputs of UAV photogrammetry are

a) DEM / DTM / DSM (surface models)
b) Orthophotos (geospatially corrected aerial images)
c) 3D Building Models
d) Contour Maps
e) Planimetric features (road edges, heights, signs, building footprints, etc.)
f) Volumetric Surveys

Some of the best photogrammetry applications using UAVs are enlisted below. The output of these systems is precision 3D images of the terrain.

- Flood Modelling
- Pollution Modelling
- Mapping and Cartography
- Urban Planning
- Coastline Management
- Transport Planning
- Oil and Gas Exploration
- Quarries and Minerals (Volumetric and Exploration)
- Archaeology
- Cellular Network Planning

The potent combination of UAV and photogrammetry (Tarhan 2011; Panigrahi 2014) can be leveraged as a low-cost technology solution in comparison to a high-cost satellite-based photogrammetric solution. Also, a benefit is increased efficiency and reduced costs compared to using traditional air or satellite survey. Also, the advantages of UAV photogrammetry can be thought of as a low cost of operation, high spatial resolution, high temporal resolution, low radiometric interference, and no need for classification of images because of the high fidelity and high spatial resolution of the acquired images. Geolocating and geocoding spatial objects are more accurate employing UAV-based photogrammetry in a small area.

The application domain of systems designed using the UAV-photogrammetry combination is emerging. Some of the areas of application are navigation, communication, geo-intelligence, surveillance, and Non-Line-of-Sight (NLoS) communication (Shakhatreh et al., 2019). The applications are being popularly used because of their real-time capability to acquire process and compute nascent spatio-temporal data acquired through the aerial platform. This provides many essential services to the user community. Some prominent services offered by an integrated UAV photogrammetry system are

- **Location services**: Locating fixed targets or moving targets (spatial objects).
- **Surveillance service**: Continuously monitoring a spatio-temporal phenomenon pertaining to the instantaneous field of view (IFoV) of the sensor (Panigrahi & Mohan, 2011; Lowe, 1999).
- **Non-Line-of-Sight (NLoS) communication service**: The drone provides an aerial platform for relaying the data from a transmitter to a receiver by providing a base platform up in the air to achieve communication between the transmitter and receiver where Line of Sight (LoS) does not exist.
- **Navigation and Surveillance services**: A UAV is essentially fitted with a receiver of a Global Navigation Satellite System (GNSS), for example, a Global Positioning System (GPS) and Inertial Navigation System (INS) for tracking and controlling the platform's aerial manoeuvres and stability in the air. This also gives the location, height from the datum surface, distance, direction, velocity, and acceleration of the moving object tracked by the UAV (Chiba et al., 2000; Yilmaz 2006).
- **Imaging and Mapping service**: The GIS (Geographic Information System) (Panigrahi 2009; Narayan 2001) gives the much-needed map services by converting the images acquired by the drone to image maps by imparting datum, map projection and coordinate system.

The key services provided by the photogrammetry-drone combination are depicted in Figure 7.1. These services provided by a photogrammetry-drone

Figure 7.1 Services provided by photogrammetry-drone.

combination have made it a useful system for many military and societal applications. Generally in the military, the operations to be performed by a drone are remote, for a large time duration, and it has to operate in the air with a small footprint or self-signature. Therefore, the key design constraint of the system is its Size, Weight, Area and Power (SWAP) (Castrillo et al., 2022). Therefore, the photogrammetric functions which coexist with the ground station need to be designed in keeping the stringent SWAP constraints for long-duration operation in an outdoor environment. Figure 7.2 depicts a typical scenario of the operation of a photogrammetry-drone combination.

Thus, designing a UAV for a photogrammetric application can be put in the form of a min-max optimization: minimization of a self-signature or electronic signature, a small but stable aerodynamic structure, less power consuming, while at the same time maximizing operation duration, payload or sensor carrying capability. The important factor of a UAV for photogrammetric application is to install a high capability software in the UAV with an AI enabled software system. The processing software can fuse the spatial data obtained from different sensors to obtain high fidelity and maximum content of spatial information. The spatial image can be classified into various natural classes using AI and deep learning algorithms to generate a category image. An on-board AI-based algorithm recognizes the spatial object of interest, and software removes geometrical error, if any, in the photogrammetric product.

Algorithms for photogrammetry typically attempt to minimize the sum of the squares of errors over the coordinates and relative displacements of the reference points. This minimization is known as bundle adjustment and is often performed using the Levenberg–Marquardt algorithm. This algorithm is now infused in the on-board computing of the UAV.

7.3 PHOTOGRAMMETRIC FUNCTIONS NEEDED FOR ISTAR

ISTAR is an acronym for Intelligence, Surveillance, Target Acquisition, and Reconnaissance. In a sense, ISTAR is a functional specification for the Observe, Orient, Decide, and Act (OODA) cycle on the battlefield. It is achieved through a system that collaborates with platforms, namely, UAV/satellite/airplane, sensors IR (Infra-Red)/LiDAR (Light Detection and Ranging)/SAR (Synthetic Radar) processes such as geometric correction, namely, Ortho-Photo-Generation/Image Registration and photogrammetric functions such as PAT (Polynomial Affine Transformation)/SLAM (Simultaneous Localization and Mapping)/Triangulation/Trilateration.

Therefore, the UAV is a natural platform for carrying out ISTAR applications which otherwise were performed by observing the battlefield using balloons, kites, manned aerial missions, or satellite surveillance. All these systems can

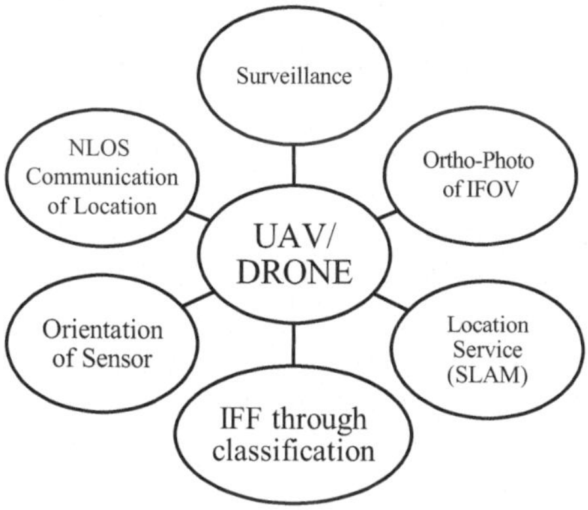

Figure 7.2 UAV as the platform for ISTAR applications.

obtain a static picture of the area under consideration and through photographic processing to identify and find the location of the object under surveillance. However, they cannot perform identification of dynamic change on the battlefield or track the location of a moving object in real time, whereas the appropriate sensor fitted with a UAV can perform ISTAR functions such as continuous surveillance of a large area of terrain, and classification of the image corresponding to Instantaneous Field of View (IFoV) of the terrain, identify objects of interest or targets, and perform Simultaneous Localization and Mapping (SLAM) of moving targets to locate and track the object. The location of the object of interest can be communicated to the weapon system to engage the target or to the field commander to take appropriate action. The ISTAR functions performed by a UAV are depicted in the bubble diagram in Figure 7.2.

The UAV has emerged as the most viable platform for ISTAR applications because it decreases the cycle time of OODA, increasing battle efficiency. The photogrammetric operations performed on board a UAV and its corresponding ISTAR functions are analysed through the discrete steps, (a)–(k), that follow:

a) Surveillance (Observe) is the operation of continuously monitoring the spatio-temporal activities of an area using sensors mounted on a static or dynamic platform, usually an aerial platform like a UAV, airplane, or satellite. The electronic footprint corresponds to the IFoV through sensors such as IR, photonics, LiDAR, or SAR.
b) This image is further rectified using bundle block correction or Polynomial Affine Transform (PAT) or similarity transform or log-polar transform to create an ortho-corrected image corresponding to the IFoV of the sensor.

c) Using the ortho-image, the geometric properties of the object such as location, area, perimeter, and volume can be measured using photogrammetric principles.
d) If the object under consideration is a moving target, the target or spatio-temporal object is located using a point inside polygon algorithm.
e) Then the target is tracked and persistently locked in the IFoV by computing the outer boundary of the object in the IFoV, then computing the Centre of Gravity (CG) of the object. The object is tracked by extrapolating or tracking the successive CG using the mean shift algorithm.
f) The orientation parameters of the sensor are computed successively so that the sensor is oriented to keep the spatial object in the middle of the IFoV of the sensor. (Orient)
g) Classification of the image obtained by the sensors into natural class using machine learning/deep learning techniques. Spatial object identification and classification of spatial objects using IFF (Identify Friend or Foe) is performed using on-board computing.
h) Using triangulation or trilateration or multilateration, the location of the object is computed. Often, other SLAM algorithms are employed to compute the location of the moving target. The location of the object is transformed from the sensor coordinate to the universal coordinate (latitude, longitude). Also, compute the grid reference corresponding to the location of the object in different precisions, namely 4/6/8/10 figure easting and northing.
i) Communicate the target locations to the ground control system or the weapon system to effectively orient and engage the weapon to the identified target. During the process of orientation of the weapon system to engage the target, crest clearance is taken into consideration. Crest clearance involves checking whether the projectile or the kinetic weapon clears the intervening crest (terrain) between the weapon and the target.
j) Once the weapon engages with the target, orient the IFoV of the UAV to the target location to take an instantaneous image of the terrain post-engagement. The image of the IFoV post-damage is compared with that of the previous IFoV after registering them using PAT. The registered pair of images is subjected to a change detection algorithm. The change detection algorithms that are used are often algebraic, where the pixel rationing, pixel differencing, regression, or change vector analysis is employed to compute the change between the multi-instantaneous images of the terrain to ascertain the Post-Damage Assessment (PDA).
k) PDA of a target is possible to ascertain instantaneously the effectiveness of the weapon. This is possible only with a drone/UAV capturing the image of the target.

The above operations of ISR is cyclic and is known as OODA loop.

7.4 DESIGN CRITERIA OF A UAV FOR PHOTOGRAMMETRIC AND ISTAR APPLICATIONS

Generally, ISTAR applications involve long and continuous surveillance of vast terrain surfaces on the battlefield. Therefore, the UAV must be designed with SWAP constraints for long-duration operations. Since it has to operate in red land, it should have stealth properties to defeat enemy air defence capabilities. It should have anti-spoofing and anti-hacking capabilities to counter the spoofing of on-board navigation sensors such as GPS and INS sensors. Also, it should have the capability to withstand enemy Electronic Counter Measure (ECM) attacks. Such a UAV is known as an Unmanned Combat Aerial Vehicle (UCAV) of class high altitude long endurance (HALE) or medium altitude long endurance (MALE). Therefore, the UAV for performing photogrammetric functions on board must be a SWAP constraint UCAV of HALE or MALE variety. It should be able to perform photogrammetric functions such as ortho-photo generation, image classification, object detection, SLAM, and terrain change detection on board the UCAV.

QUESTIONS AND ANSWERS

1. List some key photogrammetry applications using UAVs.

Answer:

- Topographic mapping: UAVs can be used to capture high-resolution aerial imagery that can be processed to create accurate topographic maps. These maps can be used for a range of applications, such as land use planning, construction, and resource management.
- 3D modelling: UAVs can be used to capture imagery from different angles to create 3D models of buildings, structures, and landscapes. These models can be used for a range of applications, such as architecture, engineering, and urban planning.
- Asset inspection: UAVs can be used to inspect assets such as power lines, wind turbines, and pipelines. High-resolution imagery captured by the UAV can be processed to detect damage or defects in the assets, which can be used for maintenance and repair planning.
- Environmental monitoring: UAVs can be used to monitor environmental conditions such as vegetation health, water quality, and air pollution. High-resolution imagery captured by the UAV can be processed to provide valuable insights into environmental conditions and changes over time.
- Disaster response: UAVs can be used to provide real-time imagery and video feeds during disaster response operations. This can help

emergency responders to assess damage, locate survivors, and plan rescue operations.

- Archaeological surveying: UAVs can be used to capture high-resolution imagery of archaeological sites, which can be processed to create 3D models and maps. This can help archaeologists to better understand the site and plan excavations.

2. What are the prime design criteria of a UAV for photogrammetry applications?

Answer: The prime design criteria of a UAV for photogrammetry applications include SWAP constraints for long-duration operation, stealth capabilities to defeat enemy air defence, anti-spoofing and anti-hacking capabilities to counter on-board navigation sensors spoofing, and the ability to withstand enemy ECM attacks. The UAV for photogrammetry must be a UCAV of HALE or MALE variety and capable of performing functions such as ortho-photo generation, image classification, object detection, SLAM, and terrain change detection on board. Therefore, the UAV must be designed to meet the specific requirements of photogrammetry applications while still being able to operate effectively in the battlefield environment.

3. What is ISTAR?

Answer: ISTAR refers to intelligence, surveillance, target acquisition, and reconnaissance, which is a functional specification for the OODA cycle on the battlefield. It is achieved through a system that collaborates with various platforms such as UAVs, satellites, and airplanes, as well as sensors like IR, LiDAR, and SAR. UAVs are a natural platform for carrying out ISTAR applications, as they can perform continuous surveillance of large areas of terrain and classify the instantaneous field of view of the terrain, identify targets of interest, and track moving targets through SLAM. The location of the object of interest can be communicated to the weapons system or field commander to take appropriate action.

4. Which photogrammetric functions are needed for ISTAR application?

Answer:

- Surveillance
- Ortho-photo of IFOV
- SKAM
- IFF through classification
- Orientation of sensor
- NLOS communication of location

5. What are the advantages of unmanned aerial vehicles (UAVs) for photogrammetry operations?

Answer:

- Cost-effective: UAVs are a cost-effective alternative to traditional manned aircraft or satellite imagery for photogrammetry operations.
- Flexibility: UAVs can be easily deployed and are highly manoeuvrable, making them ideal for capturing high-resolution aerial imagery of areas that are difficult to access or dangerous for manned missions.
- Rapid Data Acquisition: UAVs can rapidly acquire data, allowing for near-real-time processing and analysis of the imagery captured.
- High-Quality Imagery: UAVs can capture high-quality aerial imagery with high spatial resolution and accuracy, making them suitable for a wide range of photogrammetry applications.
- Remote Sensing: UAVs can be equipped with a variety of sensors, including LiDAR, hyperspectral, and thermal sensors, to capture additional data beyond just imagery.
- Reduced Risk: UAVs can significantly reduce the risk of accidents and injury associated with manned aerial missions.

6. Give a list of near-field remote sensing applications of UAV.

Answer:

- Precision Agriculture: UAVs equipped with sensors such as multispectral, hyperspectral, and thermal can help in monitoring crop health, detecting crop stress, identifying nutrient deficiencies, and optimizing irrigation.
- Infrastructure Inspection: UAVs can be used for inspecting bridges, power lines, wind turbines, and other infrastructure for damage, corrosion, and other defects.
- Disaster Management: UAVs can provide real-time information about disaster-affected areas, including damage assessment, search and rescue operations, and the monitoring of relief efforts.
- Environmental Monitoring: UAVs can be used for environmental monitoring applications such as mapping and monitoring of wildlife, land cover classification, and the monitoring of water quality.
- Mining and Quarrying: UAVs equipped with LiDAR sensors can create high-resolution maps of mining and quarrying sites for better management and planning.
- Construction and Surveying: UAVs equipped with high-resolution cameras and LiDAR sensors can be used for topographic surveys, site inspections, and construction progress monitoring.

- Forestry Management: UAVs can be used for forest management applications such as tree counting, measuring of tree height, and monitoring of forest health.
- Cultural Heritage Documentation: UAVs equipped with high-resolution cameras can be used for documenting cultural heritage sites, creating 3D models, and monitoring preservation efforts.

7. What are the design criteria for a UAV for near-field remote sensing?

Answer:

- Payload capacity: The UAV should be able to carry a payload consisting of a high-resolution camera, LiDAR sensors, thermal sensors, or other remote sensing equipment.
- Flight time: The UAV should have a long flight time to cover a large area and to allow for multiple missions in a single flight.
- Manoeuvrability: The UAV should be able to fly at low altitudes and manoeuvre in tight spaces to capture detailed images of the target area.
- Stability: The UAV should be stable and capable of maintaining a steady flight path, even in windy conditions.
- GPS accuracy: The UAV should have high GPS accuracy to enable precise georeferencing of the acquired data.
- Sensor integration: The UAV should be able to integrate various types of sensors for capturing different types of data.
- Data transmission: The UAV should be equipped with a reliable and fast data transmission system to transfer the acquired data in real time to a ground station.
- Safety features: The UAV should have safety features such as collision avoidance systems, emergency landing capabilities, and fail-safe mechanisms.
- Ease of operation: The UAV should be easy to operate, with a user-friendly interface for mission planning and control.
- Cost: The UAV should be cost-effective to acquire, operate and maintain.

REFERENCES

Aasen H., Honkavaara E., Lucieer A., Zarco-Tejada P.J. Quantitative Remote Sensing at Ultra-High Resolution with UAV Spectroscopy: A Review of Sensor Technology, Measurement Procedures, and Data Correction Workflows. *Remote Sensing*. 2018; 10(7):1091. https://doi.org/10.3390/rs10071091

Brain M. Flights of Fancy: Products in the Unmanned Systems Marketplace. *IEEE Consumer Electronics Magazine*. 2015 April; 4(2):46–48.

Castrillo V.U., Manco A., Pascarella D., Gigante G. A Review of Counter-UAS Technologies for Cooperative Defensive Teams of Drones. *Drones*. 2022; 6(3):65. https://doi.org/10.3390/drones6030065

Chiba N., Kano H., Minoh M., Yasuda M. Feature-Based Image Mosaicing. *Systems and Computers in Japan*. 2000; 31(7):1–9.

Geng L., Zhang Y.F., Wang P.F., Wang J.J., Fuh J.Y.H., Teo S.H. UAV Surveillance Mission Planning with Gimbaled Sensors. *11th IEEE International Conference on Control & Automation (ICCA)*, pp. 320–325; 2014.

Goshtasby A.A. 2D and 3D Image Registration for Medical, Remote Sensing, and Industrial Applications. Wiley Inter Science; 2005. ISBN 9780471649540.

Lowe D.G. Object Recognition from Local Scale-invariant Features. *International Conference on Computer Vision*, Corfu, Greece, pp. 1150–1157; 1999.

Mohsan S.A.H., Othman N.Q.H., Li Y., Alsharif M.H., Khan M.A. Unmanned Aerial Vehicles (UAVs): Practical Aspects, Applications, Open Challenges, Security Issues, and Future Trends. *Intelligent Service Robotics*. 2023;16(1):109–137. 10.1007/s11370-022-00452-4. Epub 2023 Jan 16. PMID: 36687780; PMCID: PMC9841964.

Panigrahi N. Computing in Geographic Information Systems. CRC Press; 2014. ISBN-9781482223149.

Panigrahi N. Geographic Information Science. Universities Press, Hyderabad; 2009. ISBN-9781482223149

Colomina, P. Mohna (2014) "Unmanned aerial system for photogrammetry and remote sensing: A review" *ISPRS Journal of Photogrammetry and Remote Sensing* 92(2014)79-97

Panigrahi N., Mohan B.K., Athithan G. Pre-Processing Algorithm for Rectification of Geometric Distortions in Satellite Images. *Defence Science Journal*. 2011 March; 61(2):174–179.

Panigrahi N., Panigrahi S.S. Processing data acquired by a DRONE using a GIS: Designing a Size-, Weight-, and Power-Constrained System. *IEEE Consumer Electronics Magazine*. 2018 March; 7(2):50–54.

Panigrahi N., Smita T. Design Criteria of a UAV for ISTAR and Remote Sensing Applications. *Journal of the Indian Society of Remote Sensing*. 49, 665–669, 2020 Oct., https://doi.org/10.1007/s12524-020-01249-7

Panigrahi N., Tripathy S. Image Registration Using Polynomial Affine Transformation. *Defence Science Journal*. 2001 July; 52:253–259.

Shakhatreh H. et al. Unmanned Aerial Vehicles (UAVs): A Survey on Civil Applications and Key Research Challenges. *IEEE Access*. 2019; 7:48572–48634. 10.1109/ACCESS.2019.2909530

Tarhan M., Altug E. A Catadioptric and Pan-Tilt-Zoom Camera Pair Object Tracking System for UAVs. *Journal of Intelligent & Robotic Systems*. Jan 2011; 61(1–4):103–118.

Yilmaz A., Javed O., Shah M. Object Tracking: A Survey. *ACM Computing Survey*. 2006 Dec; 38(4): Article 13.

Chapter 8

Autonomous Underwater Vehicles (AUVs)

OVERVIEW

This chapter introduces various aspects of Autonomous Underwater Vehicles (AUVs). The aspects of various subsystems and components constituting the AUV, and their functions are discussed. Various sensors attached to the AUV for performing sensing of the environment and navigating to the desired destination underwater are discussed. Also examined is the mechanism of communication of the AUV with its base station and other peers. Finally, different categories of the AUV as per different mission needs are discussed. As an overview, the chapter comprises the following subsections, illustrating the concepts through various diagrams.

1) What is an AUV? 2) Why is an AUV required? Application and operation overview of the AUV, 3) Subsystem overview of the AUV, 4) Various types of AUVs, 5) Classification of AUVs, 6) Sensor payload of AUVs

8.1 WHAT IS AN AUV?

The Autonomous Underwater Vehicle (AUV) is a multi-use platform constituting different subsystems and technologies. Generally, it is deployed for the collection of underwater hydrographic, geological, or environmental data for commercial or scientific exploration of large hydrographic bodies or subsurfaces of the sea (González-García et al., 2020). The AUV is also used for the exploration of seabeds for minerals and oil. In the military domain, AUVs are used increasingly as defensive measures for the protection of oil rigs, harbours, and other vital assets at sea against underwater threats. Thus, AUVs (Allard et al., 2014) are deployed for locating and clearing underwater minefields, detecting threats like torpedoes or submarines, and neutralizing external threats to shipping. AUVs are also deployed for intelligence gathering and surveillance.

In the civil domain, the AUV is used for underwater surveys for collecting bathymetric, environmental, and geological data such as depth, turbidity, sea currents, salinity, temperature, and subsurface seismic data.

DOI: 10.1201/9781003432685-8

As brought out above, the dual-use capability of the AUV has made it a highly sought-after platform for military and civil applications. Not only can an AUV perform mission objectives underwater, but also it minimizes the risk to human explorers performing these critical operations (Tadjdeh, 2018).

This section gives a brief overview of the system configuration of an AUV. It depicts pictorially the components and subsystems of the AUV (Deutsch et al., 2018). The section also discusses the configuration changes of an AUV including sensors and associated technologies for different purposes or missions. The system configuration and sensor payload of AUVs change according to the mission requirements. Therefore, different types and categories of AUVs are developed and deployed by various agencies. A brief categorization of AUVs is also presented. Further, the sensor payload required for different types of AUVs is enumerated. A scenario where a swarm of AUVs can be deployed to accomplish a mission objective is discussed to make the reading more interesting.

This section enumerates the classification of AUVs with classifying criteria such as mobility (Propelled/Guide), endurance underwater and depth of operation, weight, and payloads, mission-specific objectives, and so forth Button et al. (2009).

8.2 SYSTEMS AND COMPONENTS OF AUVS

An AUV can be considered as a system of various subsystems that are interconnected (Liu et al., 2015) and housed in a water-tight chamber, communicating with each other through electro-mechanical signals and actuations from the system's perspective. These subsystems are of different technologies complementing and supplementing themselves to perform the overall mission requirements for which the AUV is designed. A technology overview of an AUV is depicted in the bubble diagram shown in Figure 8.1.

8.2.1 Power pack

Every AUV, despite its configuration or design, requires power (generation or storage) for propulsion and operation (Deutsch et. al, 2018). The endurance of the AUV is a direct function of the capacity of the in-house power generation or storage system. Power propels the AUV to its desired destination. Also, the power determines the peak speed of movement of the AUV.

The power unit supplies the sensors to actively scan and store the data in the memory storage. Also, the communication system is required to be powered by the in-house power system to keep alive the intra- and inter-communication among the subsystems and the AUV and mother station (Bradley et al., 2001).

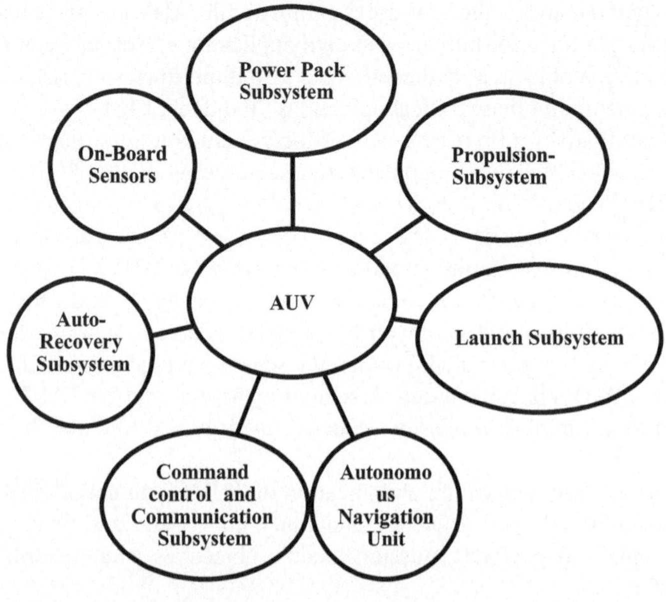

Figure 8.1 The subsystems of AUVs.

There are different power storage or generation mechanisms built into an AUV. Prominent among them are Li-ion batteries (4–5 hrs endurance) and nuclear power generation units for (1- to 30-day) operations.

8.2.2 Propeller subsystem

AUVs which are self-propelled are generally fitted with an aquatic propeller at the rear of the AUV. It displaces water to push and propel the AUV. Generally, the propulsion and navigation consume the bulk of the in-house power (Min et al., 2020).

8.2.3 Launching subsystem

Generally, AUVs are launched from the mother ship, but there are different and various mechanisms of launching AUVs (Sarda et al., 2018) into the sea subject to the surface or underwater conditions. The widely used platforms for launching an AUV are the deck of the ship using motion-compensating cranes or slipped-through cradles on the surface of the sea. Other means of launching AUVs are para-dropping or dropping into the sea by helicopter or a UAV. The

AUV is preprogrammed to manoeuvre to the desired destination as per the designated trajectory fed into the memory of the on-board computer of the AUV.

The Auto Navigation System (ANS) of the AUV takes the initial position of the AUV from the Global Positioning System (GPS) at the beginning of the mission as a seed and interpolates its position using the differential velocity and heading information obtained from the Inertial Navigation System (INS). The GPS and INS combination together constitute the ANS, which uses the dead reckoning mechanism, to estimate the current location and heading of the AUV for navigating to the destination. While estimating or computing the latitude, longitude, and depth of the AUV, the navigation system makes use of the data from various sensors included in the body of the AUV, employing a body coordinate system, and then the location is transformed to a world coordinate system using the Earth-Centered Earth-Fixed coordinate system or the coordinate system of the mother vessel. The last reported location of the AUV is often used for recovery of the AUV. The location of the AUV is often computed using the triangulation principle by obtaining the distance of the AUV from the locations of three or more buoys positioned on the surface of the sea. AUVs are also launched from the mother submarines for deep-water missions through torpedo tubes. The AUV returns to the position from where it is launched after the completion of the mission and gets docked at the launch point of the submarine and gets homed to the source from where it was launched. Generally, torpedo tube launches of AUV-form submarines are for a short duration and limited area operation. Often, the AUV surfaces to the sea surface by generating buoyancy force. After the body of the AUV is surfaced, it transmits on a specific frequency (its last location may also be transmitted) to the mother vessel for its recovery.

8.2.4 Autonomous Navigation System

One of the crucial subsystems of the AUV is the Autonomous Navigation System (ANS) (Bao et al., 2020). The ANS of the AUV is more special than that of the Unmanned Ground Vehicle (UGV) and the Unmanned Aerial Vehicle (UAV) because conventional Global Navigation Satellite Systems (GNSS) systems such as the GPS or the Global Navigation Satellite System (GLONASS) cannot be made available underwater. Therefore, the important part of the underwater operation is that AUV has to operate and navigate in a GNSS-denied situation. Radio Frequency (RF) navigation signals of the GNSS constellations do not penetrate underwater. Therefore, the ANS of the AUV has to locate itself and navigate in a GNSS-denied environment using the INS accommodated in its body and dead reckoning, or using its relative position to the buoys or beacon (Maurelli et al., 2022). Various sub-systems of an AUV and its assembly are depicted in Figure 8.2 and Figure 8.3.

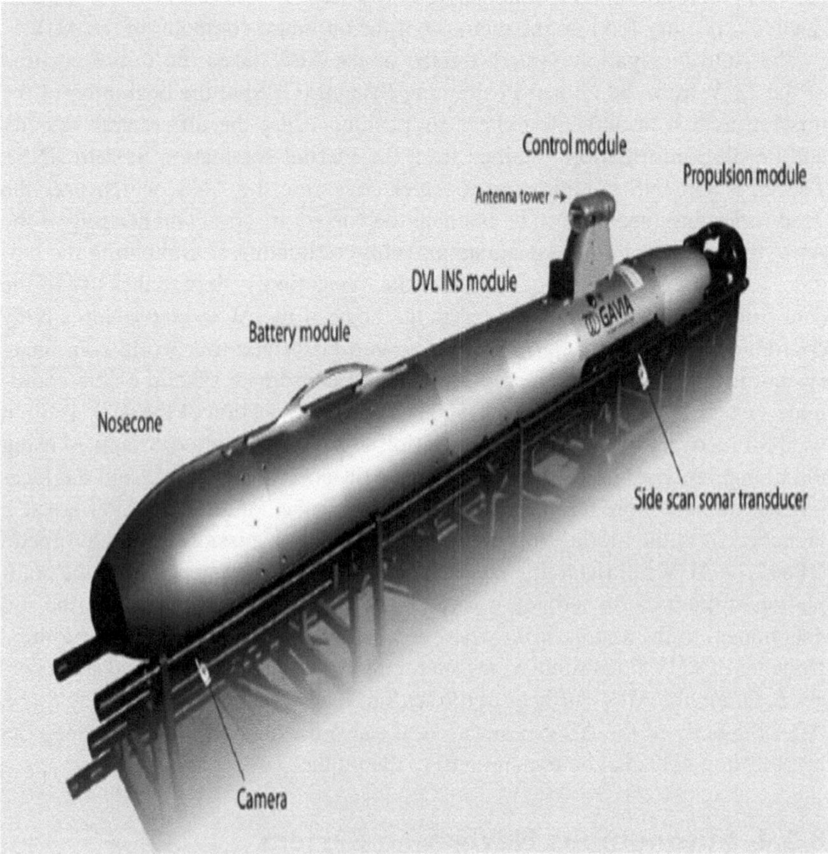

Figure 8.2 The subsystems of Underwater Vehicles.

8.2.5 Communication system

Maintaining connectivity between an AUV and the mother station is a critical aspect of the system. Due to the absence of a direct Line of Sight (LoS) and the inability of RF and other communication signals to travel long distances through water, digital communication via the sea is challenging. As a result, the AUV must use dyadic communication, opportunistic communication, or a hybrid communication method to connect and communicate with the mother vessel. In rare instances, an intermediary transmitter and receiver may act as a repeater or relay station, allowing data collected by the AUV to be transmitted to the mother station. This approach is also used to provide the mother vessel with the

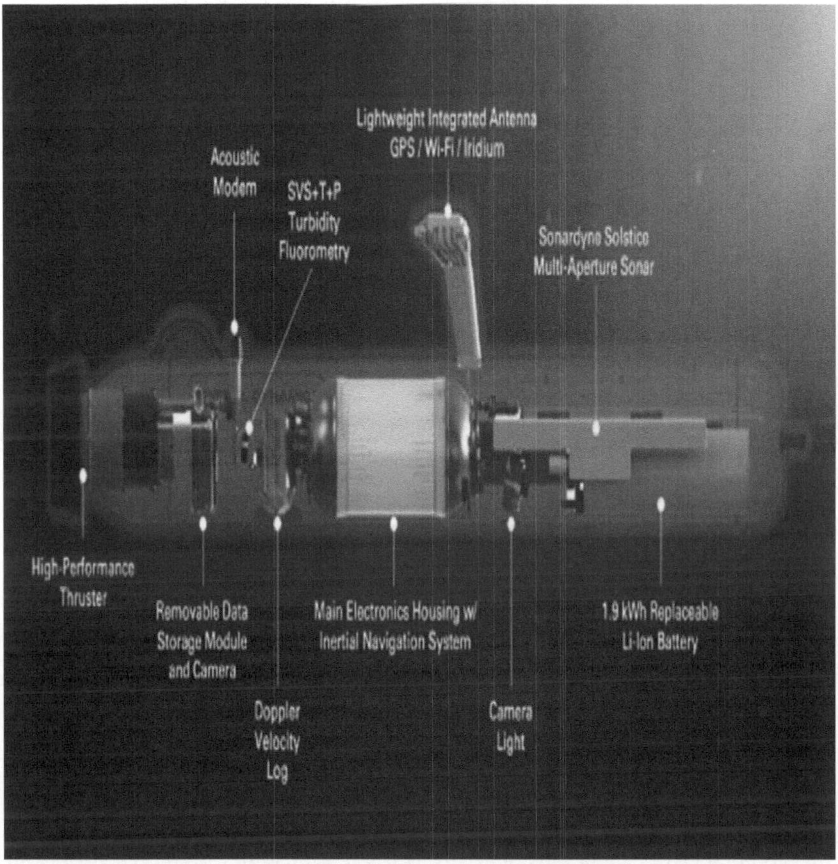

Figure 8.3 Glass-box view of subsystem assembly of an AUV.

AUV's last known location. The non-line-of-sight communication between the AUV and the mother vessel is through hybrid mediums of water, air, and mixed (last metre) using sonar, blue laser, satellite (iridium), and so forth. The mode of communication of the AUV with the mother station is depicted in Figure 8.4.

8.3 SENSOR/PAYLOAD OF THE AUV

AUVs can carry a wide variety of underwater sensors and payloads that allow them to gather data and to map from the subsurface environment (Russell et al., 2014). These may include side scan sonars, single and multi-beam sonars, magnetometers, hydrophones, and chemical sensors. Due to the AUV's inability to send or receive electromagnetic signals for a long distance when submerged,

Figure 8.4 Communication system used in AUV with the mother station through satellite, AUV communication to the data storage with sonar.

real data communication from the sensor to the platforms in the base station is difficult. Therefore, the data are stored on-board for later retrieval.

The AUV is a platform designed to carry out multiple missions. Therefore, different sensors are fitted to the AUV depending upon the requirements of the mission objectives. Generally, an AUV is fitted with some common sensors and some specific sensors. There are three broad categories of mission objectives for which AUVs are deployed (Fletcher, 2000).

a) Intelligence, Reconnaissance, and Surveillance (IRS) mission underwater
b) Anti-submarine, anti-mining operation
c) Underwater scientific exploration of hydrographic phenomena to collect hydrographic data for mapping or charting

For ISR missions, the essential sensors are sonar, depth-measuring sonar and sidewise scanning sonar, a high-resolution underwater camera, and hyperspectral and Synthetic Aperture Radar (SAR) sensor. The point echo-sounding sonar is for depth measurement, and the sidewise scanning sonar is for collision avoidance and measurement of the proximity of other foreign objects. Light Detection and Ranging (LiDAR) is for mapping the 3D topography of the sea bottom or a digital elevation model of the sea bottom surface. A high-resolution optical camera can be operated underwater for acquiring high-resolution photographs or videos of the environment. Hyperspectral and multispectral sensors (Saari et al., 2017) are deployed to acquire and classify objects of different chemical compositions.

SAR is used for scanning the sea bottom for mapping and detection of moving objects (Bovenga, 2020).

Proximity sense sensors are for collision avoidance
In addition to the above sensors, AUVs are fitted with sensors to measure turbidity, salinity, temperature, and sea current. Various sensor payloads mounted on the body of an AUV is depicted in Figure 8.5 (a) and (b) respectively.

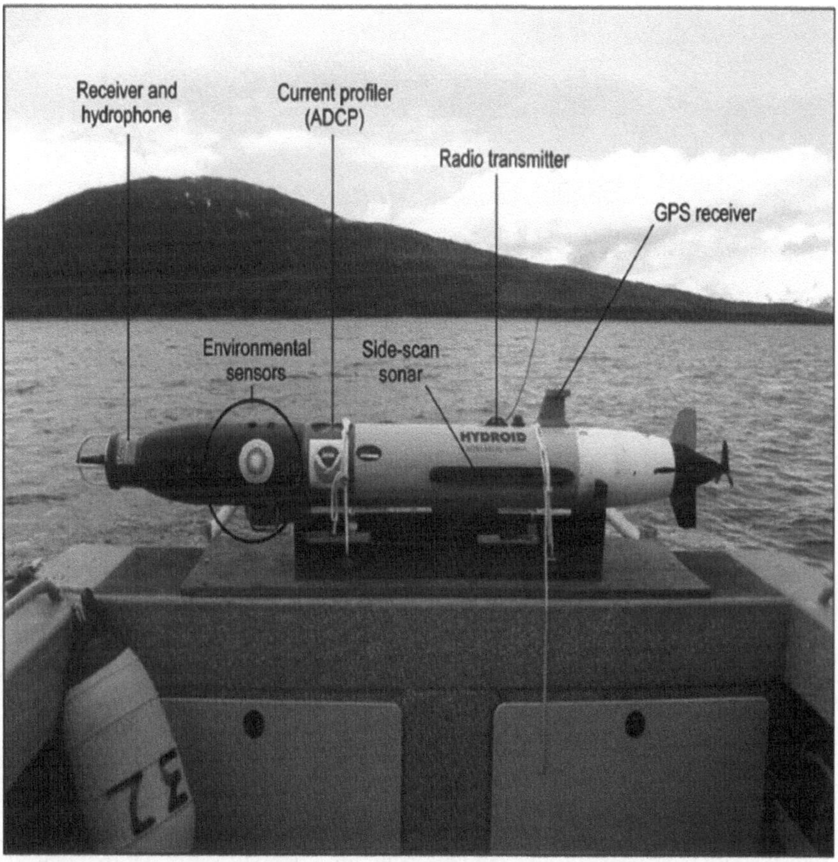

Figure 8.5 (a) Sensor payload mounted on an AUV, (b) Generic sensors mounted on the body of an AUV.

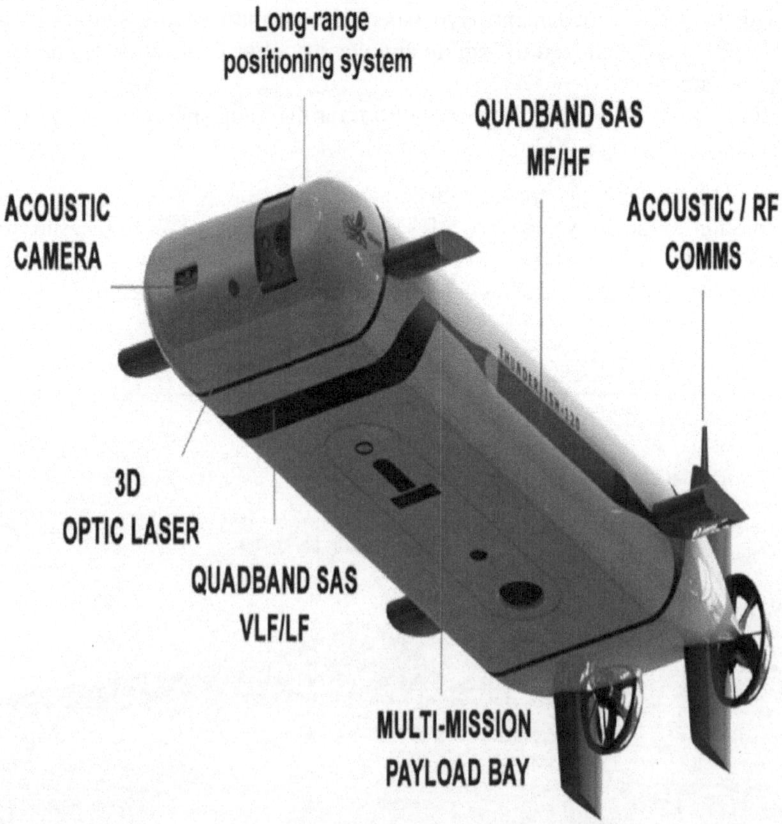

Figure 8.5 (Continued)

8.4 RECOVERY OF AUVS

Like the different mechanisms of launching, there are various recovery mechanisms of the AUV. Often, the AUV is programmed or stored with a predefined course for navigation. Therefore the start point of the mission, the path to be traversed, and the destination are stored in the on-board memory of the AUV, which is utilized by the ANS subsystem for navigation to the destination. Once the mission is completed, the AUV generates a buoyancy force in the body so that the AUV surfaces on the sea and sends its location to the mother ship or source station for recovery (Sarda et al., 2018).

The AUV, which is launched from the submarine using the torpedo chamber, comes back to the mother submarine and docks using a magnetic field through the opening of the torpedo for recovery inside the chamber.

8.5 TYPES OF UNDERWATER AND UNMANNED VEHICLES

Unmanned Underwater Vehicles (UUVs), often known as underwater drones or Uncrewed Underwater Vehicles (UUVs), are submersible vehicles that can operate underwater without a human occupant, usually categorized as Uncrewed Maritime Vehicles (UMVs) (Bajon et al, 2022). These vehicles may be divided into two categories: Remotely Operated Underwater Vehicles (ROUVs) and AUVs. ROUVs are remotely controlled by a human operator. AUVs are automated and operate independently of direct human input (Mannam et al., 2020).

The role underwater vehicles play in naval warfare cannot be overemphasized. Recently, they have been used for direct combat operations. The main reason for employing these vehicles is to explore or scout for environmental data. These systems provide clandestine and stand-off capabilities; they carry out already determined and analysed missions; and they offer a physical environment and acoustic imagery data of all relevant bodies.

UUVs, which are referred to as drone vehicles, are designed specifically to operate underwater without any human driver. These specially designed vehicles have been classified into two categories: the Autonomous Underwater Unmanned Vehicle and the remotely controlled or Supervised Underwater Unmanned Vehicle. The main difference between these two unmanned vehicles is that AUVs, as the name aptly implies, are those unmanned vehicles that operate independently, meaning that no direct human input is needed for these vehicles to function once preprogrammed. The remotely controlled or supervised unmanned vehicles are those unmanned vehicles that require human control before they can carry out their preprogrammed duty mission. These underwater vehicles have been used by naval and military organizations of various nations. These unmanned vehicles are used by these organizations not only to detect but also to destroy underwater mines.

8.6 UNMANNED UNDERWATER VEHICLES HAVE BEEN CATEGORIZED INTO FOUR CLASSES

Underwater Vehicles are classified in four distinct sub-classes as depicted in the Figure 8.6. The heavyweight class of AUVs, which have an endurance level of about 40 to 80 hours, and are launched from submarines, are one of the classes of unmanned underwater vehicles. The lightweight class is the second class, and these UUVs have an endurance level of about 20 to 40 hours. They

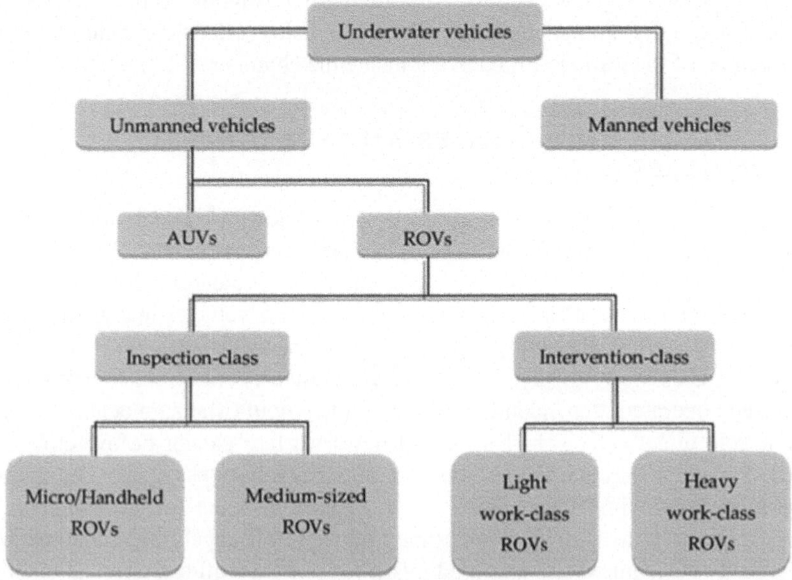

Figure 8.6 Classification of UWVs (Class Hierarchy of Underwater Vehicles).

were launched from Rigid Hull Inflatable Boats (RHIB) with the assistance of cranes or retrieval systems. The third class is the large vehicle class. Launched from submarines and surface chips, these underwater vehicles have about 10-ton displacements. The last class of these UUVs is the man-portable class. Launched by many allies from small watercraft, these UUVs have a 10- to-20-hour endurance level and a 30 to 100 lb displacement (Gafurov et al., 2015).

8.6.1 High Endurance Autonomous Underwater Vehicle (HEAUV)

The High Endurance Autonomous Underwater Vehicle (HEAUV) is a type of AUV that is designed for extended operations in the ocean. HEAUVs are typically equipped with advanced sensors, cameras, and other scientific instruments that allow them to collect data on the ocean environment over a long period of time, without the need for human intervention. Refer to Figure 8.7 for a schematic view of an HEAUV.

HEAUVs are often used in oceanographic research, exploration, and environmental monitoring. They can operate at great depths and for extended

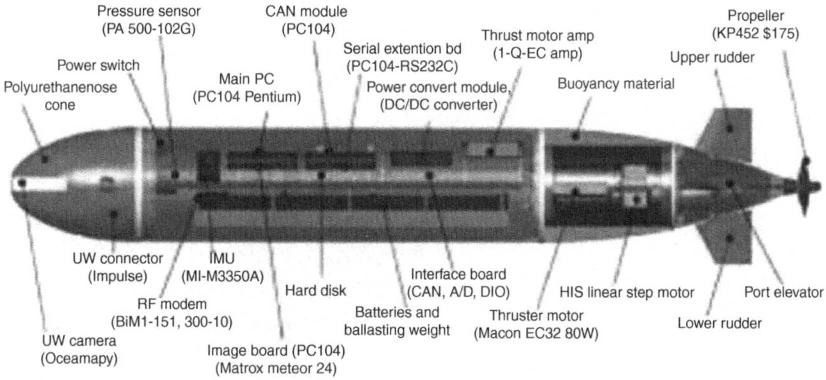

Figure 8.7 Glass box view of various components of an HEAUV.

periods, providing a persistent presence in the ocean. HEAUVs can be used for undersea mapping, pipeline inspection, and other commercial and military applications.

HEAUVs are typically powered by rechargeable batteries and can travel long distances, allowing them to cover large areas of the ocean. They are also equipped with advanced navigation and communication systems that allow them to operate autonomously, without the need for human intervention.

8.6.2 AUVs and autonomous submarines

AUVs are a subset of UUVs that can function without real-time control from a human operator (Figure 8.8). They can be programmed to follow mission way-points along a predetermined route and collect data or activate other payloads. Some autonomous submarines may use artificial intelligence capabilities to make decisions and react to changes in their environment.

8.6.3 Underwater gliders

Most AUVs are powered by rechargeable batteries, and some may also use fuel cells or solar power. The vehicles often use electric-powered propellers and thrusters for propulsion. A subset of AUVs, known as underwater gliders, use variable buoyancy to change their depth and utilize hydrofoils to convert this movement into forward motion. Underwater gliders are ideal for sampling oceanographic parameters such as temperature and salinity throughout the water column (Joo et al., 2015).

Figure 8.8 Slocum G3 Underwater Glider by Teledyne Marine.

8.7 DIFFERENT NAVIGATION METHODS USED IN AUVS

Autonomous underwater navigation (Ankit, 2021) must rely on methods other than GPS, due to the inability of radio waves to penetrate very deep into water. While the vehicle may surface from time to time to gain GPS fixes, it is required to be supplemented by other technologies as well. Marine INS can be used to gain position, acceleration, and velocity estimates, the accuracy of which can be improved with a Doppler Velocity Log (DVL), a device similar to sonar that measures the rate of travel over the sea floor, whereas depths can be measured using pressure sensors. Various hybrid navigation techniques used in AUVs are discussed in the next subsections.

8.7.1 Vision-based navigation

Subsea vehicles, particularly AUVs, have successfully relied on high-accuracy dead-reckoned INS coupled with DVLs for their positioning during underwater surveys. Unfortunately, navigation using an INS-DVL without external aiding is a methodology that inherently drifts over time, meaning error continuously increases with time during a mission. Drift can be minimized by using surface-based Ultra Short Base Line (USBL) positioning, or Long Base Line (LBL) positioning, but these solutions have limitations in deep water, and across wide areas, respectively. A truly autonomous framework will require complementary methods to deliver robust and long-duration subsea positioning (Larsen, 2000).

If we look to terrestrial applications for inspiration, to driverless cars, and to aerial drones as examples, we see their solutions to comparable problems all leverage imaging sensors and computer vision to deliver accurate navigational solutions. By detecting environmental features, the platform's relative position can be tracked Dayoub et al. (2015) while simultaneously mapping the environment as it is perceived – technically referred to as Simultaneous Localization and Mapping (SLAM). Not only does this improve reliability by aiding dead reckoning with through-the-sensor visual measurements, but it also provides a framework to reconnect with past perceptions by detecting loop closure events – an event where the autonomous vehicle recognizes a previously visited area and can correct its position with this knowledge.

It is evident that theoretically these same methodologies could be applied to underwater navigation, utilizing imaging data of a subsea environment from sensors like digital still cameras, laser scanners, or high-resolution sonar.

8.7.2 Laser loop closure

Voyis and its development partners, including the DECAR Systems Group at McGill University, have applied these terrestrial methodologies to the subsea environment by taking advantage of Voyis's high-resolution 3D laser data from its Insight laser scanners. Subsea laser scanners are uniquely suitable for this application since they generate rich quantitative data in real time, enabling the identification of environmental features during an active mission, and then detecting loop closure events when a location has been previously visited (Figure 8.9).

Loop closure algorithms were tested on a local shipwreck survey, which provides a quantifiable target to evaluate the methodology. The shipwreck was traversed on multiple survey passes to generate a sequence of loop closure events, with navigation data recorded with optional GPS aiding to provide a ground-truth trajectory. Real-time navigational data is used to identify possible loop closure regions, and then the algorithms analyse the laser data to identify and match features. Once detected, these events can be input back into the navigational solution to eliminate the accumulated error. The identified loop closure events are illustrated below, matching features both on the shipwreck and in the surrounding seabed (Figure 8.10).

The improvement to navigational positioning is demonstrated in a trajectory error graph (Figure 8.11), where the baseline INS-DVL exhibits a continuously growing error, while the laser loop closure-aided solution continually maintains an error of less than 0.1 m across the entire survey.

One benefit of this improved trajectory is drastically improved laser data registration from subsequent survey passes, utilizing loop closure feature

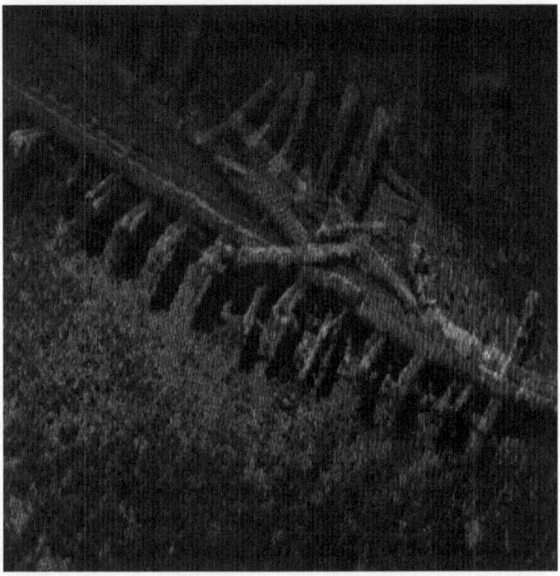

Figure 8.9 3D laser model of the shipwreck used for feature detection.

matches to align all data sets into a globally consistent model. Another benefit is to prepare a high-quality fused image of the images taken from different sensors (Figure 8.12) for image based photogrammetric measurement. This would similarly benefit other sensor data types like image-based photogrammetry, and multi-beam sonar.

8.8 FUTURE APPLICATIONS OF AUVS

The potential uses for this technology span all ocean sector survey applications. At a high level, when mapping large areas for marine research or generating digital twins of offshore wind farms and oil & gas installations, more globally consistent 3D models can be produced with improved localization of critical features. The automated data registration capability drastically improves the repeatability of accurate measurements within the dynamic laser scanning data set and simplifies the process of overlaying subsequent survey results, facilitating automated change detection analysis (Wynn et al., 2014).

When using AUVs to search for specific targets, often first identified with side-scan sonar, an optical survey that needs to deliver complete coverage can achieve consistent positioning accuracy throughout the entire mission by linking together successive survey passes, ensuring targets identified at any point in the

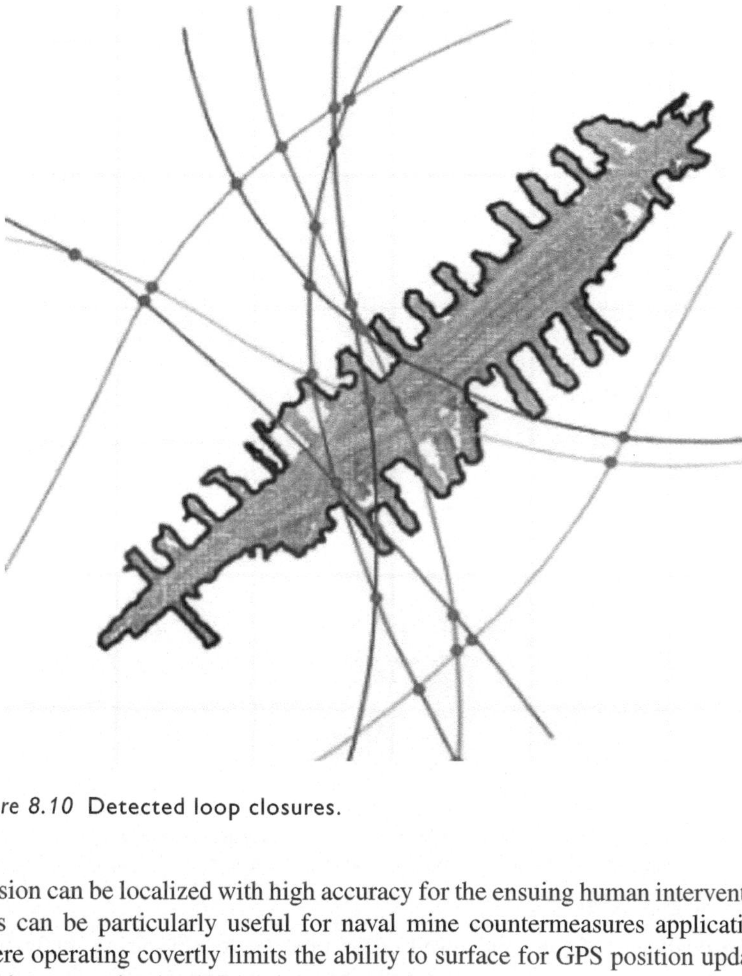

Figure 8.10 Detected loop closures.

mission can be localized with high accuracy for the ensuing human intervention. This can be particularly useful for naval mine countermeasures applications, where operating covertly limits the ability to surface for GPS position updates, or when operating in GPS-restricted locations.

Looking further ahead, this capability could be incorporated more tightly into long duration navigation methodologies, utilizing prior maps or survey databases to achieve continuous landmark navigation. Features of known locations from past surveys or existing bathymetric maps can be continually referenced, so that when the vehicle enters an established area, it can reinitialize its global position by identifying features in the area. Similarly, this capability could enhance vehicle autonomy in localized structure inspections where detected loops can support automated path planning to ensure complete inspection coverage is achieved (Jung et al,. 2008).

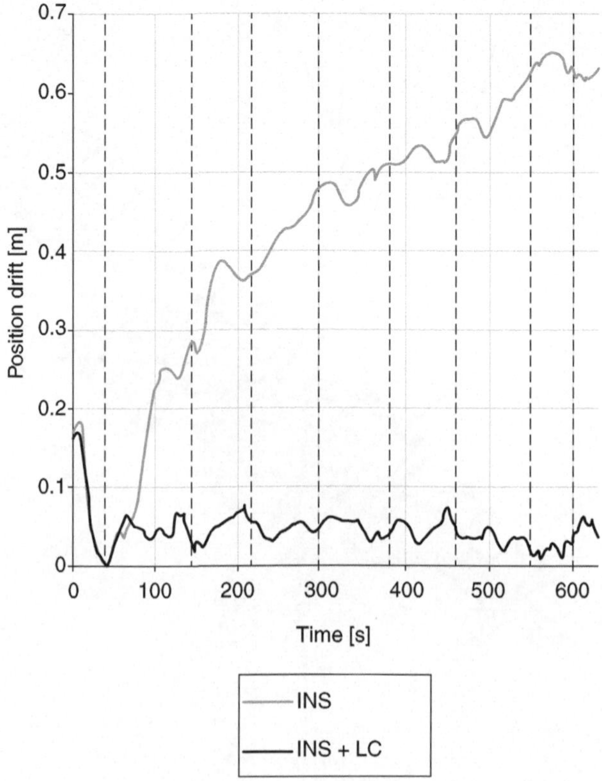

Figure 8.11 The trajectory error graph shows where the baseline INS-DVL exhibits a continuously growing error, while the laser loop closure-aided solution continually maintains an error of less than 0.1 m.

The challenge of fully uncrewed ocean operations is inspiring innovations across the ocean economy, and at both sensor and system levels, bringing us closer to a future where we can explore and understand our oceans more efficiently and with better data than ever before. The final step is putting it all together on a commercial scale to attain this vision, but true autonomy will not be achieved in isolation. It will require surveyors, vehicle designers, and sensor manufacturers all working in collaboration to illuminate the unknown

8.9 THE FUTURE OF AUVS

One problem that has plagued the use of these underwater vehicles is that of finding an adequate and efficient power supply source that will be safe to use near water. To solve this problem, the US Navy declared that before the end of

Figure 8.12 Shipwreck detected using 3D laser scanning.

2018, a drone or UUV would be developed to address this problem. It also stated that these new vehicles would be able to stay underwater for as long as 70 days.

Moreover, the US Navy is determined to develop an unmanned vehicle that will be able to perform several functions, meaning that rather than developing over 10 vehicles, they would put the time, energy, and resources into developing one that can perform multiple tasks. The US Navy is seeking to investigate and also develop an external underwater unmanned vehicle weapon.

The development and use of these Wonderful UUVs are increasing, and more companies are developing more and more, ensuring that these vehicles are functional underwater. These underwater vehicles are useful for all those involved in underwater activities.

8.10 CHALLENGES

One major issue that has stood in the way of the development and use of these unmanned vehicles is the fact that these vehicles encounter difficulty when communicating underwater. This challenge has been linked to numerous distorted transmissions and also multiple obstacles that might stand in the way of these vehicles.

Highly reliable underwater communications coupled with their low power consumption and small size have allowed advanced organizations to rapidly develop and sell solutions for various industries (Flores et al., 2020).

DSPComm provides highly reliable and versatile modems that fit the following applications:

- Collection of real-time oceanographic data
- AUV command and control
- Subsea pipeline monitoring
- Subsea telemetry control
- Submersible retrieval buoys
- River seepage monitoring
- Diver location and positioning
- Underwater asset monitoring and location-based services
- Wireless underwater lift bags
- Sample client applications

QUESTIONS AND ANSWERS

1. What is an AUV? What are the subsystems and component systems used in AUVs?

Answer: An Autonomous Underwater Vehicle (AUV) is a type of unmanned underwater vehicle that is capable of operating autonomously without human intervention. AUVs are designed for a wide range of applications, including oceanographic research, environmental monitoring, undersea mapping, and commercial and military operations.

Subsystems of an AUVs are as follows:

- Power Pack Subsystem
- Propulsion-Subsystem
- Launch Subsystem
- Autonomous Navigation Unit
- Command Control and Communication Subsystem
- Auto-Recovery Subsystem
- On-Board Sensors

2. List some scientific applications of an AUV.

Answer:

- Oceanographic research – AUVs can be used to study ocean currents, temperature, salinity, and other oceanographic parameters.
- Environmental monitoring – AUVs can be used to monitor the health of coral reefs, detect oil spills, and measure the impact of human activities on the marine environment.

- Undersea mapping – AUVs can create high-resolution maps of the seafloor and undersea features such as shipwrecks and archaeological sites.
- Marine biology – AUVs can be used to study the behaviour and distribution of marine mammals, sharks, and other marine species.
- Hydrography – AUVs can be used for bathymetric surveys, which are critical for charting the seafloor and updating nautical charts.
- Gas hydrate exploration – AUVs can be used to locate and map gas hydrate deposits, which are important potential energy resources.
- Tsunami warning – AUVs can be used to detect and monitor tsunamis, which can help provide early warning to coastal communities.
- Climate change research – AUVs can be used to study the impact of climate change on the oceans and help scientists better understand ocean–atmosphere interactions.

3. List some of the defence applications of an AUV.

Answer:

- Underwater surveillance – AUVs can be used for covert surveillance operations, such as monitoring enemy submarine activity.
- Mine countermeasures – AUVs can be used for mine detection and neutralization operations.
- Anti-submarine warfare – AUVs can be used for locating and tracking enemy submarines.
- Intelligence gathering – AUVs can be used for collecting intelligence on enemy naval operations and ship movements.
- Port and harbour security – AUVs can be used for inspecting ships and monitoring ports and harbours for security threats.
- Search and rescue – AUVs can be used for underwater search and rescue operations, such as locating downed aircraft or submerged wreckage.
- Subsea infrastructure protection – AUVs can be used for inspecting and protecting critical underwater infrastructure, such as oil and gas pipelines and communication cables.
- Underwater mapping and reconnaissance – AUVs can be used for mapping the seafloor and conducting reconnaissance missions in hostile underwater environments.

4. What is navigation? Why is navigation important in an UAV mission? How is navigation achieved in UAV missions?

Answer: Navigation refers to the process of determining the position, orientation, and velocity of a vehicle and guiding it to a desired location or along a planned trajectory.

Navigation is critical for the success of a UAV mission because the vehicle must be able to move accurately and safely through the environment while achieving its mission objectives. Without accurate navigation, the vehicle may drift off course, collide with obstacles, or fail to reach its target destination.

Navigation in UAV missions can be achieved through various means, depending on the specific application and operating environment. Some of the common methods used in UAV navigation include:

- Global Positioning System (GPS): A GPS receiver on the UAV can determine the vehicle's position by receiving signals from GPS satellites.
- Inertial Navigation System (INS): INS uses accelerometers and gyroscopes to track the UAV's movements and calculate its position, velocity, and orientation.
- Visual Navigation: UAVs can use visual cues from cameras to navigate and track landmarks.
- Radio Frequency (RF) Navigation: UAVs can use radio signals to navigate, such as employing beacon signals or triangulation.
- Acoustic Navigation: In underwater UAV missions, acoustic signals can be used for navigation, such as employing sonar to determine the distance to underwater objects.
- Hybrid Navigation: Often, a combination of the above navigation methods are used to achieve robust and accurate navigation in UAV missions.

5. What are different navigation technologies used in UAV missions?

Answer:

- Global Navigation Satellite Systems (GNSS): GNSS use signals from satellite constellations to determine the position, velocity, and time of a UAV. The most commonly used GNSS system is the US GPS, but other systems such as the Russian GLONASS, European Galileo, and Chinese BeiDou can also be used. GNSS is reliable and accurate, but it can be affected by signal interference or obstruction, especially in urban canyons or indoors.
- Inertial Navigation System (INS): INS uses accelerometers and gyroscopes to measure the UAV's motion, and from this data, estimate the position, velocity, and altitude of the vehicle. INS is generally more accurate than GNSS, but errors can accumulate over time, leading to position drift. INS can be combined with GNSS to create a hybrid navigation system that provides high accuracy and reliability.
- Vision-Based Navigation: Vision-Based Navigation uses cameras or other sensors to detect and track features in the environment, such as landmarks, roads, or buildings, and estimate the UAV's position and

velocity based on this data. This method can be used in environments where GNSS signals are weak or unreliable, such as indoors, or in urban canyons. Vision-Based Navigation requires a high level of computational power and can be affected by lighting and weather conditions.

- Radio Frequency (RF) Navigation: Radio signals, such as beacon signals or triangulation, can be used to estimate the position of a UAV.
- Acoustic Navigation: In underwater UAV missions, acoustic signals can be used for navigation, such as employing sonar to determine the distance to underwater objects.
- Light Detection and Ranging (LiDAR): LiDAR sensors can be used to create high-resolution maps of the environment, and from this data, estimate the UAV's position and velocity.

6. What is an Autonomous Navigation System? How is autonomy achieved in navigation?

Answer: An Autonomous Navigation System (ANS) is a system that enables a vehicle, such as a UAV, to navigate autonomously without human intervention. An ANS integrates various navigation sensors, such as GNSS, INS, and Vision-Based Navigation, and algorithms to enable the vehicle to plan its path and control its movements to reach its destination while avoiding obstacles and staying within operational constraints.

Autonomy in navigation is achieved through several means, including:

- Sensor Fusion: Autonomous navigation systems use multiple sensors to determine the vehicle's position, velocity, and orientation. Sensor fusion techniques are used to combine the data from these sensors to improve the accuracy and reliability of the navigation solution.
- Path Planning: Autonomous navigation systems use algorithms to plan the optimal path to reach the destination while avoiding obstacles and staying within operational constraints. These algorithms take into account the vehicle's dynamics, environmental factors, and mission requirements.
- Control Systems: Autonomous navigation systems use control systems to execute the planned path and control the vehicle's movements. These systems use feedback from sensors to adjust the vehicle's trajectory and ensure that it stays on course.
- Machine Learning: Machine learning techniques can be used to improve the performance of ANS. For example, deep learning algorithms can be used to train the system to recognize and classify objects in the environment, such as obstacles or landmarks, and adjust the vehicle's trajectory accordingly.

7. How does the ANS of an AUV differ from that of an UGV and UAV? What are some different navigational techniques used in AUV navigation in the absence of a conventional navigation mechanism?

Answer: The Autonomous Navigation System (ANS) of an AUV differs from that of a UAV and UGV due to the unique challenges posed by underwater environments. The absence of GPS signals and the limited range and bandwidth of underwater communication channels make navigation and control more challenging for AUVs than for UAVs and UGVs.

To overcome these challenges, AUVs typically use a combination of navigational techniques that can operate in the absence of a conventional navigation mechanism. Some of the common techniques used in AUV navigation include:

- Acoustic Navigation: AUVs use acoustic signals to determine their position relative to fixed transponders or acoustic beacons deployed on the seafloor. The AUV measures the time delay between transmission and reception of the signal to determine its range to the transponder or beacon. By measuring the range to multiple transponders or beacons, the AUV can determine its position.
- Inertial Navigation: AUVs use Inertial Navigation Systems (INS) that measure the vehicle's acceleration and rotation rates to determine its position and velocity. INS is especially useful in situations where GPS signals are not available, such as in underwater environments.
- Dead Reckoning: Dead reckoning is a navigation technique that involves estimating the current position of the AUV based on its previous known position and the estimated speed and direction of travel. Dead reckoning is often used in combination with other navigational techniques to improve the accuracy of the estimated position.
- Optical Navigation: Optical navigation techniques use cameras or other optical sensors to determine the vehicle's position relative to the seafloor or other landmarks. This technique can be used in combination with other navigational techniques to improve the accuracy of the estimated position.

8. How is the communication system of an AUV different from that of a UAV and UGV?

Answer:

- The communication system of a AUV is different from that of a UAV and UGV due to the unique challenges posed by underwater environments. In contrast to UAVs and UGVs, which can use wireless communication systems such as radio-frequency (RF) or cellular networks, the underwater environment severely limits the range and bandwidth of wireless communication channels.

- The communication system of an AUV typically consists of a combination of acoustic and electromagnetic communication systems. Acoustic communication is the primary means of communication for AUVs due to its ability to travel long distances underwater and penetrate through obstacles such as water columns, thermoclines, and sound speed gradients. Acoustic communication systems use low-frequency sound waves that can be transmitted over long distances and are less affected by environmental conditions. The AUVs can use acoustic signals to communicate with the surface vessel or other underwater platforms, exchange control commands, and transfer scientific data collected during the mission.
- Electromagnetic communication systems such as Radio Frequency (RF) and satellite communication are also used in some AUVs for short-range communication with surface vessels and for transferring data during brief periods when the AUV surfaces. However, electromagnetic communication systems are limited in their range and bandwidth underwater, and their signals attenuate quickly in water. Therefore, they are not used for long-range communication or real-time control of AUVs in most cases.

9. What are different sensor payloads that an AUV carries during mission? What are the specific sensors required for different missions of an AUV?

Answer:

- Navigation Sensors: Navigation sensors are essential for an AUV to navigate autonomously in the underwater environment. The navigation sensors commonly used in AUVs include:
- Inertial Measurement Unit (IMU): Measures the AUV's orientation, acceleration, and angular velocity.
- Doppler Velocity Log (DVL): Measures the AUV's velocity relative to the seafloor.
- Pressure sensor: Measures the depth of the AUV.
- Magnetic Compass: Measures the AUV's heading with respect to the Earth's magnetic field.
- Global Navigation Satellite System (GNSS): Used when the AUV surfaces to obtain position information.
- Environmental Sensors: Environmental sensors are used to measure various parameters of the underwater environment, such as temperature,

salinity, and water currents. The commonly used environmental sensors include:

- Conductivity, Temperature, Depth (CTD) Sensor: Measures temperature, salinity, and depth.
- Acoustic Doppler Current Profiler (ADCP): Measures water currents.
- Optical Backscatter Sensor: Measures the amount of suspended particles in water.
- Sonar systems: Used for bathymetric mapping, object detection, and classification.
- Payload-specific Sensors: Payload-specific sensors are sensors that are specific to a particular mission's objectives. These sensors vary depending on the mission's goals and may include:
- Cameras: Used for visual inspection, photogrammetry, and video surveys.
- Spectrometers: Used to measure water quality, chemical composition, and biological activity.
- Sampling devices: Used to collect water samples for analysis in the lab.
- Magnetometer: Used for detecting ferrous objects.

10. List some of the mission objectives of an AUV.

Answer:

- Intelligence, Reconnaissance, and Surveillance (IRS) mission underwater
- Anti-submarine, anti-mining operation
- Underwater scientific exploration of hydrographic phenomena to collect hydrographic data for mapping or charting. AUVs can be used for intelligence gathering and surveillance activities, including the detection and tracking of enemy vessels, monitoring of maritime borders, and monitoring of illegal activities such as drug trafficking.
- Mine countermeasures: AUVs equipped with sensors and detection systems can detect and neutralize underwater mines, reducing the risk to personnel and vessels.
- Anti-Submarine Warfare (ASW): AUVs equipped with sonar systems can detect and track submarines, providing valuable information to military forces.
- Underwater search and rescue: AUVs equipped with sonar and other sensors can be used to locate and assist personnel in distress or lost at sea.
- Environmental monitoring: AUVs equipped with environmental sensors can be used to monitor environmental parameters such as water

temperature, salinity, and pollution levels, providing valuable information to military forces and helping to protect the environment.

- Underwater communication: AUVs equipped with communication systems can be used to establish and maintain communication links with other military assets, including submarines and unmanned surface vessels, in areas where traditional communication methods are difficult or impossible.

11. What are different mechanisms for recovery of AUVs?

Answer: The recovery mechanism of an AUV depends on the launching method used. Typically, AUVs are programmed with a predefined course for navigation, with the start point, path, and destination stored in on-board memory and utilized by the Autonomous Navigation System (ANS) for navigation. Once the mission is completed, the AUV generates a buoyancy force in the body, causing it to surface on the sea and send its location to the mother ship or source station for recovery.

In the case of an AUV launched from a submarine using a torpedo chamber, the recovery mechanism involves the AUV returning to the mother submarine and docking using a magnetic field through the opening of the torpedo for recovery inside the chamber. This is a specialized method of recovery used for AUVs launched from submarines, where the torpedo chamber provides a secure and convenient docking location for recovery. Once the AUV docks in the chamber, it can be retrieved and serviced for its next mission.

12. What are different classes of AUVs? Give a tree diagram of different classes of AUVs with a brief note on each of the classes of AUVs.

Answer: Autonomous Underwater Vehicles (AUVs) are unmanned vehicles used for exploring and surveying the ocean. Remotely Operated Vehicles (ROVs) are robotic devices used for underwater inspection and maintenance tasks. There are various types of ROVs, including Micro ROVs, Medium-sized ROVs, Light work-class ROVs, and High Endurance Autonomous Underwater Vehicles (HEAUVs).

Micro ROVs are small and manoeuvrable, ideal for tasks such as underwater inspections and surveys. Medium-sized ROVs are more powerful and capable of carrying heavier payloads, suitable for more complex missions. Light work-class ROVs are designed for more demanding tasks such as construction and maintenance work, while HEAUVs have the ability to operate autonomously for extended periods of time, making them ideal for long-term underwater exploration and surveillance missions (Figure 8.6).

REFERENCES

Allard, Y., Shahbazian, E. (2014). Unmanned Underwater Vehicle (UUV) Information Study. Defense Technical Information Center: 78. Archived from the original on April 30, 2019.

Bajon, T., Grand-Clément, S. (2022). Uncrewed Maritime Systems: A Primer | UNIDIR. www.unidir.org. Retrieved 2022-12-07.

Bao, J., Li, D., Qiao, X., Rauschenbach, T. (2020). Integrated navigation for autonomous underwater vehicles in aquaculture: A review. *Information processing in agriculture*, 7(1):139–151.

Bovenga, F. Special Issue: Synthetic Aperture Radar (SAR) Techniques and Applications. *Sensors*. 2020; 20(7):1851. https://doi.org/10.3390/s20071851

Bradley, A., Feezor, M.D., Singh, H., Sorrell, F. (2001). Power systems for autonomous underwater vehicles. *IEEE Journal of Oceanic Engineering*. 26: 526–538. 10.1109/48.972089.

Button, R.W., Kamp, J., Curtin, T.B., Dryden, J. (2009). "A Survey of Missions for Unmanned Undersea Vehicles". *National Defense Research Institute*: 223 – via RAND.

Dayoub, F, Dunbabin, M, Corke, P. (2015). Robotic Detection and Tracking of Crown-of-Thorns Starfish. In *2015 IEEE/RSJ International Conference on Intelligent Robots and Systems (IROS)* (pp. 1921-1928). IEEE.

Deutsch, C., Thune, S., Soderling, F., Moratelli, L., Kuttenkeuler, J. (2018). Design of an AUV Research Platform for Demonstration of Novel Technologies. 10.1109/AUV.2018.8729729.

Fletcher, B. (2000). New roles for UUVs in intelligence surveillance and reconnaissance. *Proceeding MTS*.

Flores, H., Hossein Motlagh, N., Zuniga, A., Liyanage, M., Passananti, M., Tarkoma, S., Youssef, M., Nurmi, P. (2020). Toward Large-Scale Autonomous Monitoring and Sensing of Underwater Pollutants. *arXiv preprint arXiv:2005.09571.*

Gafurov, S., Klochkov, E. (2015). Autonomous Unmanned Underwater Vehicles Development Tendencies. *Procedia Engineering*. 106. 10.1016/j.proeng.2015.06.017.

Goel, A., Paredes, J.A., Dadhaniya, H., Ul Islam, S.A., Mohammed Salim, A., Ravela, S., Bernstein, D. (2021). Experimental Implementation of an Adaptive Digital Autopilot. *2021 American Control Conference (ACC)*, pp. 3737–3742.

Gonzalez-Garcia, J., Gomez-Espinosa, A., Cuan-Urquizo, E., Garcia-Valdovinos, L.G., Salgado-Jimenez, T., Cabello, J.A.E. (2020). Autonomous Underwater Vehicles: Localization, Navigation, and Communication for Collaborative Missions. *Applied Sciences*. 10(4):1256. https://doi.org/10.3390/app10041256

Joo, M.G., Qu, Z. (2015). An Autonomous Underwater Vehicle as an Underwater Glider and Its Depth Control. *International Journal of Control, Automation and Systems*. 13, 1212–1220. https://doi.org/10.1007/s12555-014-0252-8

Jung, Y.-S., Lee, K., Lee, B. (2008). Advances in Sea Coverage Methods Using Autonomous Underwater Vehicles (AUVs). 10.5772/5478.

Larsen, M.B. (2000). High Performance Doppler-Inertial Navigation-Experimental Results. *OCEANS 2000 MTS/IEEE Conference and Exhibition. Conference Proceedings (Cat. No.00CH37158), Providence, RI*, vol. 2, pp. 1449–1456. 10.1109/OCEANS.2000.881808.

Liu, H., Shao, J., Zhou, Q., Joseph-Ellison, S., Jaworski, J., Wen, C., Shi, X. (2015). The Mechatronic System of Eco-Dolphin – a Fleet of Autonomous Underwater Vehicles. 10.1109/ICAMechS.2015.7287138.

Mannam, P., Mahbub Alam, M., Krishnankutty, P. (2020). Review of Biomimetic Flexible Flapping Foil Propulsion Systems on Different Planetary Bodies. *Results in Engineering*. 8:100183. 10.1016/j.rineng.2020.100183.

Maurelli, F., Krupinski, S., Xiang, X., Petillot, Y. (2022). AUV Localisation: A Review of Passive and Active Techniques. *International Journal of Intelligent Robotics and Applications*. 6. 10.1007/s41315-021-00215-x.

Min, F., Pan, G., Xu, X. (2020). Modeling of Autonomous Underwater Vehicles with Multi-Propellers Based on Maximum Likelihood Method. *Journal of Marine Science and Engineering*. 8(6):407. https://doi.org/10.3390/jmse8060407

Wynn, R.B., Huvenne, V.A.I., Le Bas, T.P., Murton, B.J., Connelly, D.P., Bett, B.J., Ruhl, H.A., Morris, K.J., Peakall, J., Parsons, D.R., Sumner, E.J., Darby, S.E., Dorrell, R.M., Hunt, J.E. (2014). Autonomous Underwater Vehicles (AUVs): Their Past, Present And Future Contributions to the Advancement of Marine Geoscience. *Marine Geology*. 352: 451–468. ISSN 0025-3227.

Saari, H., Akujarvi, A., Holmlund, C., Ojanen, H., Kaivosoja, J., Nissinen, A., Niemeläinen, O. (2017). Visible, Very Near IR and Short Wave IT Hyperspectral Drone Imaging System for Agriculture and Natural Water Applications. *ISPRS International Archives of the Photogrammetry, Remote Sensing and Spatial Information Sciences*. 165–170.

Sarda, E., Dhanak, M. (2018). Launch and Recovery of an Autonomous Underwater Vehicle From a Station-Keeping Unmanned Surface Vehicle. *IEEE Journal of Oceanic Engineering*. 1–10. 10.1109/JOE.2018.2867988.

Tadjdeh, Y. (2018). Annual Naval Exercise Showcases Unmanned Underwater Vehicle Capabilities. *National Defense*. 103(780):24–26. JSTOR 27022380. Retrieved 2022-09-02.

Chapter 9

Essential design concepts of AUVs

OVERVIEW

To design an Autonomous Underwater Vehicle (AUV), it is essential to have background knowledge and fundamental concepts about the processes and physical laws governing the AUV in its environment (Mohd Aras et al., 2009). Experimental studies have confirmed that the factors governing the operation of an AUV underwater are buoyancy, hydrodynamic damping, and pressure of the water, which have to be taken into consideration. The stability of the AUV depends upon its geometry and the speed of its motion underwater.

This section introduces some of these fundamental concepts and ideas about the stability of the moving geometry and kinematics of underwater vehicles and also examines the general design principles of these vehicles. The mechanical and electrical systems of AUV vehicles are also examined closely to gain insight into different design aspects.

9.1 AUV COORDINATE SYSTEM

Analogous to Unmanned Aerial Vehicles (UAVs) or any other flying vehicles, an underwater vehicle has six DOF (Degree of Freedom); three spatial coordinates, x, y, and z; and three attitudes defining Euler angles, roll, Ø, pitch, θ, and yaw, ψ, namely (x, y, z, Ø, θ, ψ) (Bae et al., 2014) as depicted in Figure 9.1)

The X axis points along the forward direction of the vehicle, defining the vehicle's longitudinal transition. The Y axis points through the right-hand or horizontal side of the vehicle, defining the vehicle's lateral transition. The Z axis defines the vehicle's vertical transition or depth transition. It is assumed and noted that the value of z coordinates is zero at the sea surface and points downwards; hence, it is positive for increasing depth. This is because the Autonomous Underwater Vehicle (AUV) cannot travel farther than the surface of the water.

The degrees of freedom that an underwater vehicle allows are quite versatile. This high range of motion, however, can be problematic as it must be considered during the design process how each degree of freedom will be controlled to

DOI: 10.1201/9781003432685-9

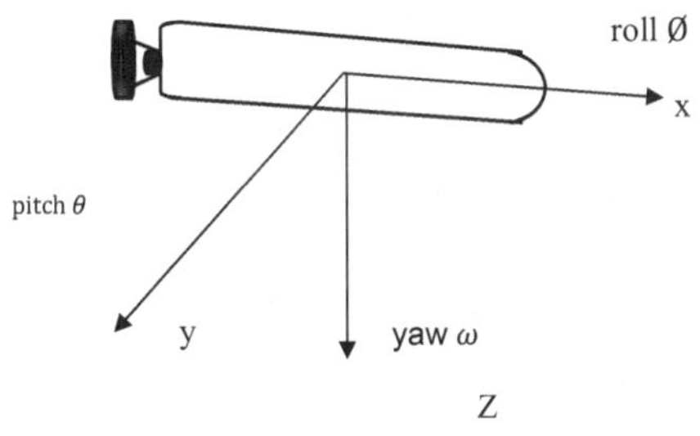

Figure 9.1 Free body diagram of different forces acting on an AUV.

keep the vehicle stable. Most AUVs are designed to be able to control as many degrees of freedom as possible.

9.2 FACTORS AFFECTING AN UNDERWATER VEHICLE

Several forces act on an underwater vehicles that affect the kinematics of the vehicle, and they require careful consideration for the design process. These include buoyancy, hydrodynamic damping, Coriolis, and added mass. Buoyancy is one of the most important factors which significantly affects the vehicle's ability to submerge as well as its stability while in motion. Stability is also affected by external forces such as sea current and pressure of the water column, which are other significant factors for underwater vehicles that need to be taken into consideration in the design process (Chen et al., 2021).

9.3 BUOYANCY

The magnitude of the buoyant force, B, exerted on a body, floating or submerged, is equal to the weight of the volume of water displaced by that body. The ability of an object to float depends on whether or not the magnitude of the weight of the body, W, is greater than the buoyant force B. As per Archimedes' principle, if (B > W), then the body will float, whereas if (B < W), then it will sink (Ayyangar et al., 2018). This is depicted in Figure 9.2. If (B == W), then the body remains where it is.

9.4 HYDRODYNAMIC DAMPING

When a body is moving through the water, the main forces acting in the opposite direction to the motion of the body are hydrodynamic damping forces. These damping forces are mainly due to drag and lifting forces, as well as lineal skin friction Damping forces have a significant effect on the dynamics of an

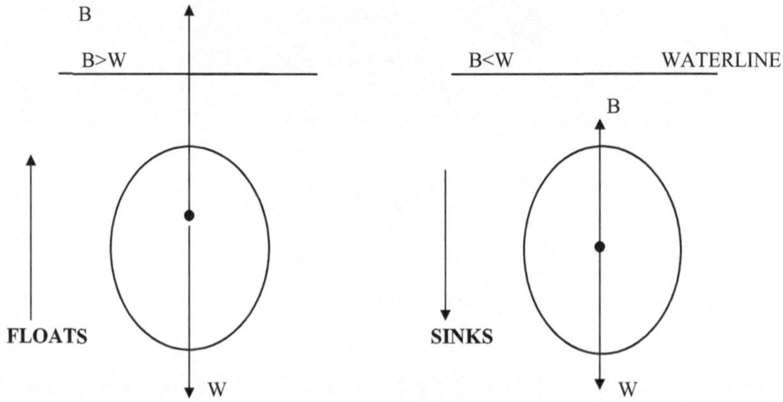

Figure 9.2 Conditions for the AUV to remain floating.

underwater vehicle, which leads to nonlinearity (Gafurov and Klochkov, 2015). Lineal skin friction can be considered negligible when compared to drag forces, and therefore, it is usually sufficient to take into account only the latter drag and lifting forces when calculating damping forces (Kim et al., 2002).

9.5 STABILITY OF AUVS

Assuming no water movement, the stability of a static body underwater is predominantly affected by the positions of the centres of mass, $C_{M,}$ and buoyancy, C_B (Elnashar, 2013). The centre of buoyancy is the centroid of the volumetric displacement of water due to the structure of the body. If C_M and C_B are not aligned vertically with each other in either the longitudinal or the lateral direction, there will be instability due to the creation of a non-zero moment. This is depicted in Figure 9.3.

If C_M and C_B coincide in the same position in space, the vehicle will be very susceptible to perturbations. Ideally, the two centroids should be aligned vertically some distance apart from each other with C_M below C_B. This results in an ideal bottom-heavy configuration with the innate stability of the structure.

As seen in Figure 9.4, this configuration produces a righting momentum, RM, when the vehicle rolls or pitches, that is directly proportional to the perpendicular distance between C_M and C_B as well as to both B and W. This moment is conducive to the vehicle's stability, acting as a passive roll and pitch control system. The moment is given by,

$$RM = \frac{1}{2} d \left(B + W \right), \qquad 9.1$$

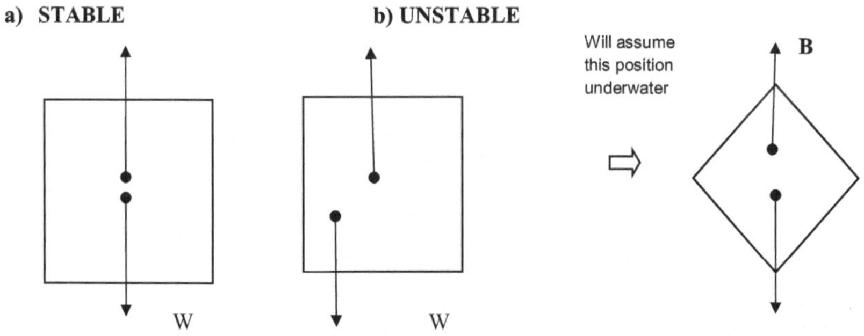

Figure 9.3 a) Stable configuration of the underwater body, b) Instability of an underwater body through misalignment of centres of mass and buoyancy.

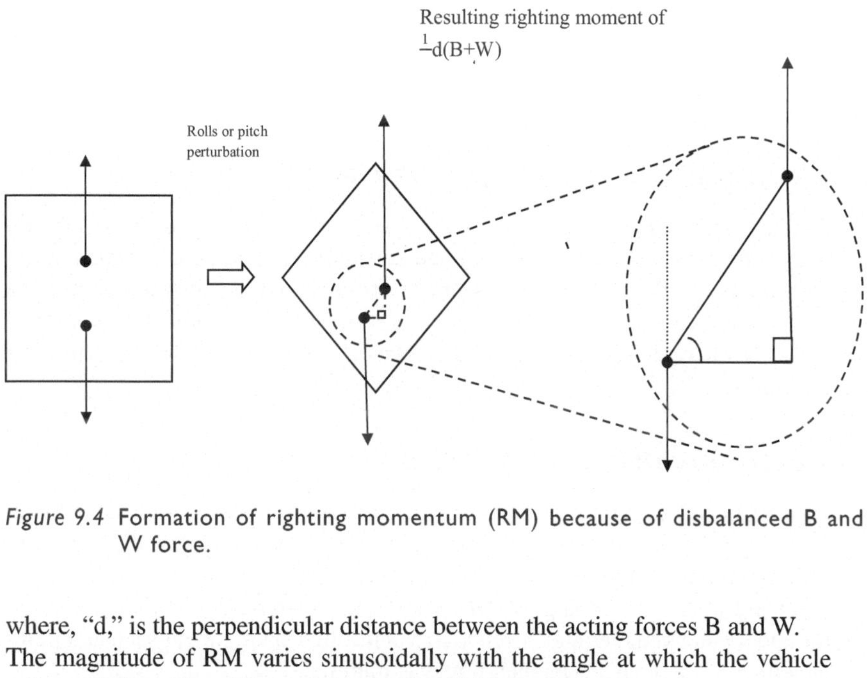

Figure 9.4 Formation of righting momentum (RM) because of disbalanced B and W force.

where, "d," is the perpendicular distance between the acting forces B and W. The magnitude of RM varies sinusoidally with the angle at which the vehicle rolls or pitches. From Figure 9.4, equation 9.1 becomes,

$$\text{RM}=\frac{1}{2}l(B+W)\sin\lambda, \qquad 9.2$$

where L, B, and W, are constants for the vehicle; l being the distance between the centres of mass and buoyancy.

In the case of a dynamic underwater body, stability is affected not only by the centres of mass and buoyancy but also by factors such as external forces and centres of drag. To increase dynamic stability, the centres of drag, determined by the centroids of the effective surface areas of the vehicle, should be aligned with the centres of the externally applied forces. In this manner, the vehicle will not tend to exhibit undesirable characteristics in its motion.

9.6 CORIOLIS

Coriolis is an inertial force that acts perpendicular to the direction of motion of the body. The force is proportional to both the velocity and rotation of the coordinate system. The effect of the Coriolis force, then, is that the path of the body is deflected. In reality, however, the path of the body is not actually deflected, but only appears to be. This is due to the motion of the body's coordinate system. Since the coordinate system of a AUV rotates with respect to another reference frame, the effect of the Coriolis force is usually taken into account and included in the equation of motion (Aruna, 2022).

9.7 ADDED MASS DUE TO MOTION

Another phenomenon that affects underwater vehicles is added mass. When a body moves underwater, the immediate surrounding fluid is accelerated along with the body. This affects the dynamics of the vehicle in such a way that the force required to accelerate the water can be modelled as an added mass. Added mass is a fairly significant effect and is related to the mass and inertial values of the vehicle.

9.8 ENVIRONMENTAL FORCES

Environmental disturbances can affect the motion and stability of the vehicle. This is particularly true for an underwater vehicle where waves, currents, and even wind can perturb the vehicle. When the vehicle is submerged, the effect of wind and waves can be largely ignored. The most significant disturbances then for underwater vehicles are currents. In a controlled environment such as a pool, the effect of these environmental forces is minimal.

9.8.1 Pressure

As with air, underwater pressure is caused by the weight of the medium, in this case, water, acting upon a surface. Pressure is usually measured as an absolute or ambient pressure, absolute denoting the total pressure and ambient being of a relativistic nature.

At sea level, pressure due to air is 14.7 psi or 1 atm. For every 10 m of depth, pressure increases by about 1 atm, and hence, the absolute pressure at 10 m underwater is 2 atm. Although linear in nature, the increase in pressure is significant, and underwater vehicles must be structurally capable of withstanding a relatively large amount of pressure if they have to survive in deep sea conditions.

9.9 GENERAL DESIGN OF AN AUV

There are several aspects of AUVs, such as its mechanical and electrical design, that need to be looked at closely. International submarine engineering identifies hull design, propulsion, submerging, and electric power as the major four aspects of designing an AUV (Saghafi and Lavimi, 2020).

9.9.1 Hull design

An AUV must provide a pressure hull to house its components in a dry, watertight environment. The hull must allow components to be easily accessible and maintainable, as well as allow for modularity in case of future changes or additions. As well as being light and strong, the hull should also be corrosion-resistant as it will be subjected to a harsh saltwater environment (Nouri et al., 2015).

Spherical hulls offer the best structural integrity; however, the shape inhibits the efficient use of the space available as most components and systems are rectangular in shape. Cylindrical hulls provide the best alternative, comprising high structural integrity and a shape conducive to the housing of electronic components.

9.9.2 Propulsion design

Some sort of propulsion is required in all AUVs and is usually one of the main sources of power consumption. Most AUVs use motors for propulsion due to the scarcity and cost of alternative systems (Allotta et al., 2015).

The location of the motors affects which degrees of freedom can be controlled. The positioning of the motors can also affect noise with on-board electronic components, as well as propeller-to-hull and propeller-to-propeller interactions. Propeller-to-hull and propeller-to-propeller interactions can have unwanted effects on the dynamics of an AUV.

When travelling at a constant speed, the thrust produced by the motors is equal to the friction or drag of the vehicle, that is,

$$\text{Thrust} = \text{Drag} = \frac{1}{2}\rho S^2 A C_D, \quad 9.3$$

where Ais the water density, s is the speed, A is the effective surface area, and C_D is the drag coefficient

Power consumption for the propulsion system increases dramatically as the speed of the vehicle increases. This is because the thrust power is equal to the product of the thrust and the speed, meaning thrust power is a function of speed cubed,

$$\text{Thrust Power} = \text{Thrust X s} = \frac{1}{2}\rho S^3 A C_D \quad 9.4$$

Therefore, because of an AUV's limited energy supply, it must travel at a speed that does not draw too much power, but at the same time not take too long to complete its mission. Obtaining the ideal speed becomes an optimization problem.

9.9.3 Submerging

In the case of a submersible vehicle, since the volume of the vehicle remains constant, to dive deeper, it must increase the downward force acting upon it to counteract the buoyant force. It can accomplish this either by increasing its mass via the use of ballast tanks or by using external thrusters (Tan et al., 2020).

Ballasting is the more common approach for submerging. This method is mostly mechanical in nature and involves employing pumps and compressed air to take in and remove water. The alternative is to use thrusters that point downwards. This is a much simpler system, but is quite inefficient in terms of power consumption and not suited at great depths.

To reduce the size of ballast tanks or the force required by thrusters for the process of submerging, AUVs are usually designed to have residual buoyancy. That is, the weight of the vehicle is made to be more or less equal to the buoyant force.

9.9.4 Electric power

Electric power is commonly provided via sealed batteries. The ideal arrangement of batteries is to have them connected in parallel with diodes between each one to allow even discharge and to prevent current flow between batteries. Fuses or other protective devices (Bradley et al., 2001).

QUESTIONS AND ANSWERS

1. Give a free-body diagram describing various forces acting on an Autonomous Underwater Vehicle (AUV).

Answer: Refer Figure 9.1.

2. What are the different motions the AUV undergoes during its mission?

Answer:

- Pitch: The motion of the AUV around its lateral axis is known as pitch.
- Roll: The motion of the AUV around its longitudinal axis is known as roll.
- Yaw: The motion of the AUV around its vertical axis is known as yaw.
- Surge: The motion of the AUV along its longitudinal axis is known as surge.
- Sway: The motion of the AUV along its lateral axis is known as sway.
- Heave: The motion of the AUV along its vertical axis is known as heave.
- Hover: The motion of the AUV at a fixed location in the water column without any horizontal or vertical motion is known as hover.

3. What are the different forces acting on an AUV during its mission?

Answer:

- Gravity force: The weight of the AUV caused by the Earth's gravitational field.
- Buoyancy force: The force exerted on the AUV by the water that allows it to float or sink in the water.
- Hydrodynamic forces: The forces exerted on the AUV due to the movement of water around it, including drag and lift forces.
- Thrust force: The force generated by the AUV's propulsion system to move it forward or change its direction.
- Coriolis force: The force exerted on the AUV due to the rotation of the Earth, which can affect its navigation.
- Magnetic force: The force exerted on the AUV due to the Earth's magnetic field, which is used for navigation in some cases.

4. What is buoyancy force? How is the buoyancy force acting against the weight of the AUV to keep it floating?

Answer:

Buoyancy force is the upward force acting on an object submerged in a fluid, which is caused by the difference in pressure between the top and bottom of the object. It is the force that enables objects to float in a fluid.

In the case of an AUV, the buoyancy force acts against the weight of the AUV to keep it afloat. This is achieved by having the overall density of the AUV be less than that of the surrounding water. This can be accomplished by using materials with low density or by using ballast tanks that can be filled with water to increase the density and sink the AUV, and then emptied to decrease the density and allow the AUV to surface. By controlling the buoyancy force, an AUV can be made to float, dive, or maintain a particular depth in the water.

5. What is the condition for an AUV to be stable during its mission?

Answer:

For an AUV to be stable during its mission, it needs to be neutrally buoyant, which means that the weight of the vehicle should be equal to the buoyancy force acting on it. In addition, the centre of gravity of the AUV should be aligned with the centre of buoyancy, and the vehicle should have a streamlined shape to minimize drag and turbulence. The control system of the AUV should also be designed to maintain stability by adjusting the vehicle's attitude and depth as necessary.

6. What are the environmental forces acting on an AUV?

Answer: Refer Figure 10.2.

- Hydrostatic pressure: The pressure exerted by the water on the AUV increases with depth and can affect the stability of the AUV.
- Coriolis force: This force is caused by the rotation of the Earth and can affect the motion of the AUV.
- Ocean currents: The movement of water in the ocean due to currents can affect the speed and direction of the AUV.
- Waves: The motion of waves on the surface of the water can cause the AUV to experience rolling and pitching motions.
- Tides: The rise and fall of sea levels due to the gravitational forces of the moon and sun can affect the depth at which the AUV operates.

7. Describe the generic designing aspects of an AUV.

Answer:

- Shape: AUVs typically have a streamlined shape to reduce drag and increase manoeuvrability. The shape may also be optimized for specific tasks, such as surveying, mapping, or reconnaissance.

- Propulsion: AUVs use various methods of propulsion such as propellers, thrusters, or fins. These are typically powered by batteries or fuel cells. The choice of propulsion system is dependent on the mission requirements and the operating environment.
- Sensors and Payloads: AUVs are equipped with various sensors and payloads, such as cameras, sonars, magnetometers, and water quality sensors, depending on the mission requirements. The placement of these sensors is carefully designed to optimize the sensing capabilities of the AUV.
- Power Source: AUVs require a power source to run their propulsion systems and on-board electronics. The power source is typically a rechargeable battery, although some AUVs use fuel cells or other power sources.
- Materials: AUVs are typically constructed from materials such as aluminium, carbon fibre, or composite materials that are lightweight, strong, and corrosion-resistant.
- Control System: AUVs are controlled by an autonomous navigation system (ANS) that uses inputs from sensors and mission objectives to navigate the AUV along a predetermined path.
- Communication: AUVs may be equipped with acoustic or radio communication systems to transmit data to a surface vessel or shore station.
- Recovery Mechanism: AUVs require a recovery mechanism that enables them to be retrieved after completing their mission. This may involve the use of a winch, crane, or other lifting mechanism.
- The hull design of an AUV is crucial for ensuring hydrodynamic efficiency and manoeuvrability. It should be designed to reduce drag and resistance, and also provide enough space to house the payload, sensors, and propulsion systems. The hull also needs to be robust and able to withstand the harsh underwater environment.
- The propulsion system of an AUV needs to be efficient, reliable, and provide adequate thrust for the desired speed and manoeuvrability. Depending on the mission requirements, the propulsion system can be electric, hydraulic, or a combination of both. The AUV may also have different types of thrusters for different motions, such as vertical, forward, and lateral.
- The submerging design of an AUV includes the ballasting system and the control surfaces. The ballasting system is responsible for controlling the buoyancy of the AUV and maintaining its desired depth. The control surfaces include the fins and rudder, which help in controlling the AUV's motion and direction underwater. The AUV may also have sensors to detect the water pressure and adjust the ballasting system accordingly.

8. What are the sensors used on the surface of an AUV and what are their functions?

Answer:

- Global Positioning System (GPS): GPS is used to determine the precise location of the AUV.
- Inertial Measurement Unit (IMU): IMU is used to measure the orientation and motion of the AUV.
- Magnetometer: A magnetometer is used to measure the strength and direction of the Earth's magnetic field. This information is used by the ANS (Autonomous Navigation System) to determine the AUV's heading.
- Pressure sensor: A pressure sensor is used to measure the water pressure at the AUV's current depth. This information can be used to calculate the AUV's depth and altitude.
- Temperature sensor: A temperature sensor is used to measure the water temperature around the AUV. This information can be used for a variety of purposes, such as detecting thermal layers in the water column.
- Salinity sensor: A salinity sensor is used to measure the salinity of the water around the AUV. This information can be used to detect differences in water density, which can affect the AUV's buoyancy.
- Camera: A camera can be used to capture images and video of the AUV's surroundings. This information can be used for scientific research or to aid in navigation.

9. What is Coriolis force and how does it affect the motion and kinematics of an AUV during a mission?

Answer:

Coriolis force is an inertial or fictitious force that acts on objects in motion within a frame of reference that rotates with respect to an inertial frame. In a reference frame with clockwise rotation, the force acts to the left of the motion of the object. In one with anticlockwise (or counterclockwise) rotation, the force acts to the right. Deflection of an object due to the Coriolis force is called the Coriolis effect.

Coriolis force is an apparent force that arises due to the rotation of the Earth. It deflects the motion of an object moving in a straight line on the Earth's surface, causing it to follow a curved path. In the context of AUVs, Coriolis force affects the motion and kinematics of the vehicle during its mission. As the AUV moves through the water, the Coriolis force causes it to experience a lateral deflection, which can affect the accuracy of its navigation and positioning. This can be particularly important for long-duration missions where the accumulation of small errors can lead to significant deviations from the desired course. To mitigate the effects of Coriolis force, AUVs are often equipped with advanced navigation and positioning systems that take into account the vehicle's motion and the effects of environmental forces such as Coriolis force.

10. What is buoyance force? How does it differ in different depths of the water column? How buoyancy is used in the AUV mission?

Answer: Buoyancy force is the upward force that a fluid exerts on an object that is partially or fully submerged in it. It is equal to the weight of the fluid that the object displaces. The buoyancy force acting on an AUV is used to offset the weight of the vehicle and keep it afloat.

The buoyancy force acting on an AUV differs at different depths of the water column because the density of seawater changes with depth. As the depth increases, the pressure and temperature of seawater increase, which results in a corresponding increase in density. This means that the buoyancy force acting on an AUV at a particular depth will be different from that at a different depth.

To control the buoyancy of an AUV during a mission, the vehicle may use a variety of techniques, including adjusting its internal ballast, using adjustable buoyancy devices, or manipulating the volume of its flotation materials. These techniques enable the AUV to control its depth and attitude as required for the mission.

REFERENCES

Allotta, B., Costanzi, R., Gelli, J., Pugi, L., & Ridolfi, A. (2015). Design of a modular propulsion system for MARTA AUV. *OCEANS 2015 – Genova*. https://doi.org/10.1109/OCEANS-GENOVA.2015.7271397

Aras, M., Shahrieel, M., Kasdirin, H., Jamaluddin, M., & Basar, M. (2009). Design and Development of an Autonomous Underwater Vehicle (AUV-FKEUTeM). Proceedings of MUCEET2009 Malaysian Technical Universities Conference on Engineering and Technology, MUCEET2009, MS Garden, Kuantan, Pahang, Malaysia.

Aruna, M.V. (2022). Mathematical Modeling and Stability Analysis of an Effective Design of Biomimetic AUV. *Journal of Intelligent & Robotic Systems*, 106, 75. https://doi.org/10.1007/s10846-022-01768-0

Ayyangar, V.B.S., Krishnankutty, P., Korulla, M., & Panigrahi, P.K. (2018). Stability Analysis of a Positively Buoyant Underwater Vehicle in Vertical Plane for a Level Flight at Varying Buoyancy, BG and Speeds. *Ocean Engineering*, 148, 331–348. doi: 10.1016/j.oceaneng.2017.11.030. ISSN 0029-8018.

Bae, S., Shin, D.H., Kwon, S., & Joo, M. (2014). An LQR Controller for Autonomous Underwater Vehicle. *Journal of Institute of Control, Robotics and Systems*, 20. 10.5302/J.ICROS.2014.13.9005.

Bradley, A., Feezor, M.D., Singh, H., & Sorrell, F. (2001). Power Systems for Autonomous Underwater Vehicles. *IEEE Journal of Oceanic Engineering*, 26, 526–538. 10.1109/48.972089.

Chen, X., Bose, N., Brito, M., Khan, F., Thanyamanta, B., & Zou, T. (2021). A Review of Risk Analysis Research for the Operations of Autonomous Underwater Vehicles, *Reliability Engineering & System Safety*, 216, 108011. ISSN 0951-8320. https://doi.org/10.1016/j.ress.2021.108011.

Elnashar, G. (2013). Performance and Stability Analysis of an Autonomous Underwater Vehicle Guidance and Control. 2013 Proceedings of International Conference on Modelling, Identification and Control, ICMIC 2013, pp. 67–73.

Gafurov, S.A., & Klochkov, E.V. (2015). Autonomous Unmanned Underwater Vehicles Development Tendencies. Procedia Engineering. Proceedings of the 2nd International Conference on Dynamics and Vibroacoustics of Machines (DVM2014) September 15–17, 2014 Samara, Russia, vol. 106, pp. 141–148. ISSN 1877-7058. 10.1016/j.proeng.2015.06.017.

Kim, J., Kim, K., Choi, H. S., Seong, W., & Lee, K.-Y. (2002). Estimation of Hydrodynamic Coefficients for an AUV Using Nonlinear Observers. *IEEE Journal of Oceanic Engineering*, 27(4), 830–840. https://doi.org/10.1109/JOE.2002.805098

Nouri, N.M., Zeinali, M., & Jahangardy, Y. (2015). AUV Hull Shape Design Based on Desired Pressure Distribution. *Journal of Marine Science and Technology*, 21(2), 203–215. https://doi.org/10.1007/S00773-015-0343-0

Saghafi, M., & Lavimi, R. (2020). Optimal Design of Nose and Tail of an Autonomous under Water Vehicle Hull to Reduce Drag Force Using Numerical Simulation. *Proceedings of Institution of Mechanical Engineers, Part M, Journal of Engineering for the Maritime Environment*, 234(1), 76–88.

Tan, L., Huy, T., & Pham, H.-H. (2020). A Measure of Control of Submerging and Surfacing of an Autonomous Underwater Vehicle for Exploring Underwater of Rivers. *Science & Technology Development Journal – Engineering and Technology*, 3, SI151–SI156. 10.32508/stdjet.v3iSI1.756.

Chapter 10

Unmanned Ground Vehicles (UGVs)

OVERVIEW

One of the critical technological advancements which will exert a decisive role in all future surface communication and warfare is the widespread deployment of unmanned systems in surface communication and combat situations. This is particularly relevant in the domain of Unmanned Ground Vehicles (UGVs), wherein commercial, tactical, and strategic advantages can be leveraged in surface communication and especially in battle against adversaries.

UGVs have been widely researched, developed, and deployed by many countries and agencies worldwide both in civil transport and on the battlefront. Many research groups worldwide are using applied research to design and develop ways to produce a UGV or modify an existing vehicle or Battle Tank (BT) or BMP platform into a UGV.

Technologically an UGV can be thought of as a conglomeration of four subsystems, namely a physical moving platform with autonomous localization, perception, and an Autonomous Navigation System (ANS). The moving platform can be a tracked or wheeled vehicle depending upon the surface condition of a motorable road or a cross-country non-motorable terrain surface. The localization system is a set of sensors mounted suitably on the body of the vehicle to continuously sense and compute the location of the vehicle in the form of latitude, longitude, altitude, and time using a Global Navigation Satellite System and Inertial Navigation System combined.

The perception module gathers the current situation and senses the environment to acquire the terrain conditions, obstacles, and other impediments to safe navigation and decides the next direction of movement of the vehicle, keeping in view the objective to reach the destination location.

An ANS combined with the Drive-by-Wire of the UGV will enable the UGV to navigate long distances from source to destination without human intervention. This capability is expected to help reduce the cognitive load and stress on the manned crew during transportation in case of civil communication or battle crew in the case of a combat vehicle operation. It can also lead to a reduction in the number of crew members required to operate the UGV.

DOI: 10.1201/9781003432685-10

Designing an ANS require suitable identification of sensors for the ANS and mounting them in appropriate locations on the vehicle to sense and perceive the location of the UGV continuously and actuate the movement of the UGV to the next location through an electromechanical means in such a manner so as to optimally reduce the cost function.

10.1 BRIEF HISTORY OF DEVELOPMENT OF UGVS

A working remote-controlled car was reported in the October 1921 issue of RCA's *World Wide Wireless* magazine. The car was unmanned and controlled wirelessly via radio; it was thought the technology could someday be adapted to tanks. In the 1930s, the USSR developed the tele-tank, a machine gun-armed tank remotely controlled by radio from another tank. These were used in the Winter War (1939–1940) against Finland and at the start of the Eastern Front after Germany invaded the USSR in 1941. During World War II, the British developed a radio control version of their Matilda II infantry tank in 1941 (Singer, 2009a, b). Known as "Black Prince," it would have been used for drawing the fire of concealed anti-tank guns, or for demolition missions. Due to the costs of converting the transmission system of the tank to Wilson-type gearboxes, an order for 60 tanks was cancelled.

From 1942, the Germans used the Goliath tracked mine for remote demolition work. The Goliath was a small tracked vehicle carrying 60 kg of explosive charge directed through a control cable. Their inspiration was a miniature French tracked vehicle which the Germans found after defeating France in 1940. The combination of higher cost, low speed, reliance on a cable for control, and poor protection against weapons meant it was not considered a success (Douglas et al., 1995).

The first major mobile robot development effort named Shakey was created during the 1960s as a research study for the Defense Advanced Research Projects Agency (DARPA) (Yang et al, 2015). Shakey was a wheeled platform that had a television camera, sensors, and a computer to help guide its navigational tasks of picking up wooden blocks and placing them in certain areas based on commands.

10.2 WHAT IS A UGV?

An Unmanned Ground Vehicle (UGV) is a vehicle that operates while in contact with the ground and without an on-board human presence. UGVs can be used for many applications where it may be inconvenient, dangerous, or impossible to have a human operator present. Generally, the vehicle will have a set of sensors to observe the environment, and will either autonomously make decisions about its behaviour or pass the information to a human operator at a different location who will control the vehicle through teleoperation. The UGV is the land-based counterpart to Unmanned Aerial Vehicles and Unmanned Underwater Vehicles. Unmanned robotics are being actively developed for both civilian and military use to perform a variety of dull, dirty, and dangerous activities.

10.3 WHAT ARE THE DIFFERENT USES OF A UGV?

Military UGVs have a wide range of uses and can be quickly reconfigured to perform multiple roles. One of the primary uses of military UGVs is Explosive Ordnance Disposal (EOD), where they are used to safely locate and dispose of explosive devices. UGVs can also be used to carry equipment and supplies, reducing the burden on soldiers and increasing their mobility. There is great variation in designing a UGV for military role versus for a civilian application (Pekka et al., 2010).

In addition to these logistical roles, military UGVs can be used for forward reconnaissance, providing real-time intelligence and surveillance data to commanders (Maheswaran and Murugesan, 2020). They can also be equipped with weapons and used as mobile weapons platforms, providing additional firepower and support to ground troops.

Another important use of military UGVs is in manned–unmanned teaming, where they work alongside human soldiers to enhance their capabilities and effectiveness. For example, a UGV could be used to provide cover or support fire while soldiers advance on a position.

Overall, military UGVs are versatile and adaptable platforms that can perform a wide range of roles in support of military operations.

10.4 COMPONENTS AND SUBSYSTEMS OF UGVS

Based on its application, UGVs will generally include the following components:

- Physical platform
- Sensors
- Navigation, guidance, and control systems
- Communication links
- Mission-specific payloads
- Systems integration features such as Graphical User Interfaces (GUI) and remote control consoles.

10.4.1 Platform

The platform is the physical foundation of the UGV, and its design (Pekka et al., 2010) is a crucial factor in determining the UGV's capabilities and performance. The choice of locomotion system is an important consideration when designing the platform, as it determines the UGV's ability to traverse different types of terrain and obstacles. Tracks are ideal for rough, off-road terrain, as they provide good traction and stability, while wheels are better suited for smooth, paved surfaces. Legged locomotion, inspired by animals like insects and quadrupeds, can enable the UGV to traverse extremely challenging terrains such as stairs and rubble piles.

The articulated body consisting of modular units can allow for greater flexibility and versatility in the UGV's movement and interaction with its environment. This is particularly useful in scenarios where the UGV needs to navigate tight spaces or work in confined areas. For example, a snake robot could be used to explore the interior of a collapsed building or a pipeline.

The Drive-by-Wire (DBW) system, which allows for complete computer control of the UGV's movement, enables greater precision and autonomy in the UGV's actions. This can be particularly useful in scenarios where the UGV needs to operate in hazardous environments, such as minefields or chemical spills. The use of a DBW system can also reduce the physical strain on human operators, allowing them to focus on higher-level tasks such as decision making and situational awareness.

10.4.2 Sensors

The primary purpose of UGV sensors is to enable the UGV to perceive and understand its environment. This includes determining its location, detecting and avoiding obstacles, and navigating through its operational area. To achieve this, UGVs are equipped with a range of sensors that provide real-time data on the UGV's surroundings.

Some commonly used sensors in UGVs include GNSS (GPS, IRNSS, etc.) receivers for determining the UGV's location, Inertial Measurement Units (IMUs) such as magnetometers, accelerometers, and gyroscopes for measuring the UGV's orientation and movement, and odometers and inclinometers for measuring distance travelled and changes in elevation.

UGVs also use a variety of sensors to detect and avoid obstacles in their path. These include colour/IR/thermal cameras for visual perception, scanning laser range finders (LiDAR) for high-resolution 3D mapping, ultrasound range finders for short-range obstacle detection, RADAR for long-range obstacle detection, and depth cameras for 3D perception.

By combining data from these different sensors, the UGV is able to build a detailed and accurate picture of its surroundings and navigate safely and effectively through its operational area.

10.4.3 Navigation, guidance and control systems

Navigation, guidance, and control systems are essential components of UGVs that enable them to operate in a wide range of environments and perform various tasks. Navigation refers to the UGV's ability to determine its current location and to plan a path to a destination location while avoiding obstacles in its environment. This is achieved through the use of sensors, such as GPS receivers, IMUs, and range sensors, which provide information on the UGV's position, orientation, and distance to nearby obstacles. Guidance involves the UGV's ability to follow the planned path using suitable commands to the DBW system. This requires accurate and reliable control algorithms that take into account the UGV's kinematics and dynamics, as well as environmental factors such as terrain, slope, and weather conditions. To achieve this, the guidance system must be integrated with the navigation and control systems to ensure smooth and stable operation.

Control refers to the UGV's ability to execute the commands generated by the guidance system using actuation devices such as motors, brakes, and steering mechanisms. Control algorithms must take into account the UGV's physical characteristics, such as mass, inertia, and tire friction, and must be able to compensate for disturbances caused by external factors such as wind gusts and rough terrain. In autonomous UGVs, navigation, guidance, and control are all implemented in software using Artificial Intelligence (AI) algorithms. These algorithms must be carefully designed and tuned to specific UGV platforms and applications to ensure optimal performance and robustness in a wide range of environments and operating conditions.

In teleoperated UGVs, the navigation, guidance (Velaskar et al., 2014), and control systems are typically controlled by a remote human operator using a joystick, keyboard, or other input device. In this case, the UGV's systems must be designed to provide the operator with the necessary feedback and situational awareness to enable safe and effective operation.

10.4.4 Communication links

Communication links are crucial for the operation of UGVs, as they allow for the transmission of data and control signals between the UGV and its operator. The type of communication link used depends on the application and the environment in which the UGV is operating.

For short-range line-of-sight (LoS) operations, wired links such as ethernet and fibre optic connections can be used. These links provide high data throughput and low latency, making them suitable for tasks that require real-time feedback and control, such as teleoperated UGVs. Short-range wireless links such as Wi-Fi, Bluetooth, and Zigbee can also be used, but their data throughput and range are generally lower than wired connections.

For long-range, non-line-of-sight (NLoS) operations, specialized wireless links such as VHF and satellite radios are required. These links can provide extended ranges, but their data throughput and latency may be higher than that of short-range links. Therefore, they are more suitable for applications that do not require real-time feedback and control, such as autonomous UGVs.

In choosing a communication link, the range, latency, and data throughput should be carefully considered. For example, if the UGV is being teleoperated, the operator typically requires a video feed at a minimum of 15 frames per second and must be able to send control commands to the UGV at an equivalent rate, with a latency of around 25 milliseconds. In some scenarios, it may also be necessary to provide intermediate repeater stations to extend the range of the communication link.

10.4.5 Mission-specific payloads

Mission-specific payloads are a critical component of UGVs, as they enable the UGV to perform its intended tasks and achieve its objectives. The specific payloads used by a UGV will depend on its intended application and mission requirements. For example, in surveillance missions (Maheswaran and Murugesan, 2020), the payload may consist of long-range, high-resolution colour and thermal cameras mounted on a controllable pan-tilt unit. This allows the UGV to provide all-around, long-distance views and gather real-time intelligence data (Ryu et al., 2005, p. 1).

In bomb disposal applications, the payload may include a manipulator arm that can be remotely controlled to defuse or detonate a suspected bomb. Chemical sensors may also be used to detect remotely the presence of explosives and other hazardous materials. UGVs used for armed first response may be equipped with remotely controlled guns or other weapons systems as part of their payload. This allows the UGV to provide additional firepower and support to ground troops in combat situations.

Thus mission-specific payloads are an essential part of UGV design and enable the UGV to perform a wide range of tasks in support of its intended mission.

10.4.6 Systems integration features

One of the most important aspects of UGVs is the operator interface, also known as the Human-Robot Interface (HRI) or Human-Machine Interface (HMI). This interface enables human operators to interact with the UGV and control its actions, often in stressful or challenging conditions. To make the UGV easy to operate, the HRI should be designed with usability and intuitiveness in mind. This can include teleoperation devices such as joysticks or car-like driving consoles that mimic familiar control interfaces. Point-and-click graphical user interfaces can also be used to provide operators with a visual representation of the UGV and its environment, allowing them to control all aspects of the UGV's operation.

Voice-activated interfaces are another option for controlling UGVs. These interfaces allow operators to issue commands to the UGV using natural language, reducing the need for manual input and increasing efficiency. Overall, the design of the HRI is a critical aspect of UGV systems integration and can have a significant impact on the ease-of-use and effectiveness of the UGV.

10.5 KIND OF SENSORS USED IN AN UGV?

Sensors are used in the UGV for localization (knowing where the UGV is) and perception (knowing what is around the UGV, in terms of obstacles and free space for movement) (Nowakowski et al., 2023).

GNSS receivers are typically used to estimate the current location of the UGV operating outdoors. These are used together with inertial sensors like accelerometers and gyroscopes to provide better accuracy and to estimate the position and orientation of the UGV even when GNSS signals are absent or suffer from poor quality. For UGVs operating indoors, other techniques such as odometry and other sensors such as Inertial Navigation Systems, LiDAR, and cameras are used to help the UGV determine its position through a technique called Simultaneous Localization And Mapping (SLAM) (Xiaoji et al., 2017, p. 1).

Several sensors are available for perception, and the choice of sensors is made based on the application. Normally, a combination of sensors is required, since no one sensor can provide all the required information in all the operating conditions. For example, colour cameras provide rich texture and colour information of the scene, but need good lighting conditions and do not provide good quantitative (depth) information. Thermal cameras work well at night, but are less effective during the day. LiDAR provides very good quantitative information (3D range measurements of the scene) and work in day and night conditions, but are expensive. Cameras and LiDAR also perform poorly in bad weather conditions such as heavy rain and dense fog. Radars work well in all weather conditions, but have poorer resolution (Figure 10.1).

10.6 HOW DOES AN UNMANNED GROUND VEHICLE (UGV) WORK?

A UGV is a type of mobile robot that operates on land without a human operator. UGVs can be remotely controlled via a handheld or fixed control station, or operate autonomously using advanced sensors and algorithms. The degree of human intervention in the operation of a UGV can vary. Remotely operated UGVs are fully controlled by human operators, while fully autonomous UGVs require zero human intervention. There are also intermediate platforms that can perform some tasks autonomously and others under human control (Figure 10.2(a)).

Figure 10.1 A UGV with sensors mounted on its body.

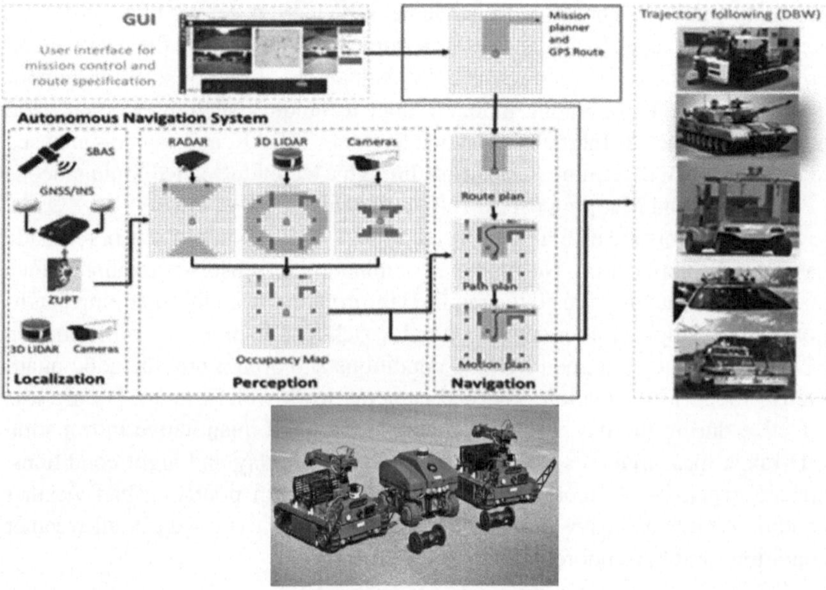

Figure 10.2 a) Flow of Action of an UGV, b) Various form factors of a mini-UGV.

Autonomous UGVs are typically equipped with advanced sensors and algorithms that enable them to navigate their environment and perform tasks without human intervention (Figure 10.2(b)). For example, an autonomous UGV may be programmed to travel between predefined waypoints while avoiding obstacles, or explore its environment to build a map. Tasks that involve a high degree of risk to life and materiel, such as weapons firing, are typically performed under human control with adequate safety interlocks. This ensures

Figure 10.3 A tracked UGV.

that the UGV operates safely and effectively while minimizing the risk of accidents or damage.

Overall, UGVs are versatile and adaptable platforms that can operate in a wide range of environments and perform a variety of tasks, either autonomously or under human control (Figure 10.3).

10.7 APPLICATIONS OF UGVS

There are a wide variety of UGVs in use today (Figure 10.3(a)(b)). Predominantly, these vehicles are used to replace humans in hazardous situations, such as handling explosives and in bomb-disabling vehicles, where additional strength or smaller size is needed, or where humans cannot easily go. Military applications include surveillance, reconnaissance, and target acquisition. They are also used in industries such as agriculture, mining, and construction. UGVs are highly effective in naval operations; they have great importance in helping Marine Corps combat; they can additionally avail in logistics operations on land and afloat (Guruprasad et al., 2020, p. 3028). The Figure 10.4 depicts the Control and Command flow chart of a UGV operations.

UGVs are also being developed for peacekeeping operations, ground surveillance, gatekeeper/checkpoint operations, and urban street presence, and to enhance police and military raids in urban settings. UGVs can "draw first fire" from insurgents – reducing military and police casualties. Furthermore, UGVs are now being used in rescue and recovery mission and were first used to find survivors following 9/11 at Ground Zero (Gavhane & Ansari, 2013, p. 4).

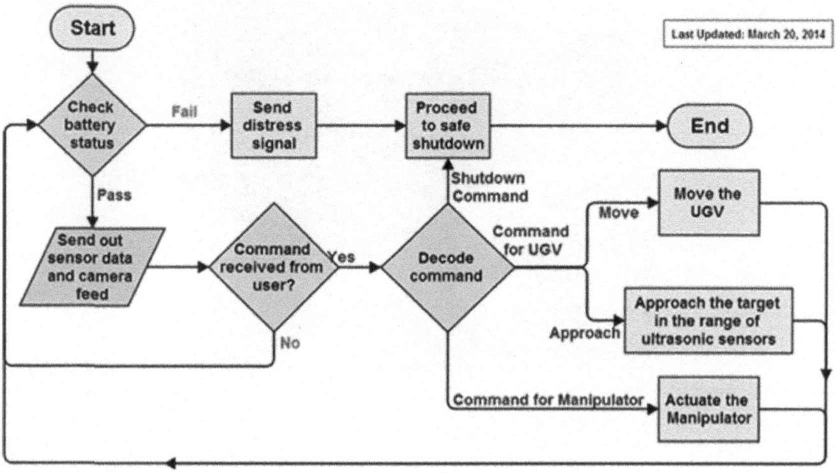

Figure 10.4 Control and command flow chart of a UGV.

10.8 AUTONOMOUS VEHICLES FOR SPACE EXPLORATION

The success of NASA's Mars Exploration Rover project is a testament to the capability of autonomous vehicles in space exploration. Both *Spirit* and *Opportunity* rovers were designed to operate autonomously on the Martian surface, travelling long distances over harsh terrain, collecting scientific data, and transmitting it back to Earth. These rovers were equipped with a suite of sensors, cameras, and scientific instruments that enabled them to explore the Martian environment and conduct detailed analyses of the geology and mineralogy of the planet.

The ability of these UGVs to operate autonomously in a harsh and challenging environment such as Mars is a significant achievement in the field of space exploration. The rovers were programmed to make their own decisions about where to go and what to investigate, based on preprogrammed scientific objectives and constraints. They were also equipped with redundant systems to ensure that they could continue to operate even if individual systems failed.

The success of the Mars Exploration Rover project has paved the way for further exploration of our solar system and beyond using autonomous UGVs. In addition to Mars, autonomous vehicles have been used for missions to other celestial bodies, such as the Moon and Saturn's moon, Titan. These vehicles have enabled scientists to conduct detailed analyses of these bodies and gather valuable scientific data, which would not have been possible using human explorers alone.

QUESTIONS AND ANSWERS

1. What is a UGV? Give some popular examples of UGVs used in societal applications.

Answer: An Unmanned Ground Vehicle (UGV) is a vehicle that operates while in contact with the ground and without an on-board human operator. UGVs are used in a variety of societal applications, including transportation, delivery, agriculture, and construction. One popular example of a UGV used in societal applications is autonomous cars. These vehicles use sensors and software to navigate roads and highways without human intervention. They have the potential to reduce traffic congestion, improve safety, and increase mobility for people who cannot drive.

Another example of a UGV used in societal applications is delivery robots. These robots are designed to deliver packages and goods to customers without the need for human drivers. They can navigate sidewalks and streets using sensors and GPS technology.

In agriculture, UGVs are used as autonomous tractors or harvesters that can work around the clock without human intervention. These machines can increase efficiency and reduce labour costs for farmers.

Finally, UGVs are also used in construction for tasks such as excavation or demolition. These machines can operate in hazardous environments where it may be unsafe for humans to work.

2. Give a bibliographic description of the development of UGV.

Answer: The development of Unmanned Ground Vehicles (UGVs) can be traced back to the early twentieth century when a remote-controlled car was reported in RCA's *World Wide Wireless* magazine in October 1921. In the 1930s, the USSR developed tele-tanks, which were machine gun-armed tanks remotely controlled by radio from another tank. The British developed a radio control version of their Matilda II infantry tank in 1941, known as "Black Prince," for drawing the fire of concealed anti-tank guns or for demolition missions. During World War II, the Germans used the Goliath-tracked mine for remote demolition work. However, due to cost, low speed, reliance on a cable for control, and poor protection against weapons, it was not considered a success. The first major mobile robot development effort named Shakey was created during the 1960s as a research study for the Defense Advanced Research Projects Agency (DARPA), which was subsequently developed.

3. What are different components of an UGV? Describe different functions of various components of an UGV.

Answer: The different components of an UGV include:

1. Physical platform: This includes the chassis, wheels or tracks, and other mechanical components that make up the vehicle.
2. Sensors: These are used to gather information about the environment around the UGV. Examples of sensors include cameras, LiDAR, radar, and sonar.
3. Navigation, guidance, and control systems: These systems are responsible for controlling the movement of the UGV. They use information from sensors to determine the vehicle's location and orientation.
4. Communication links: These are used to transmit data between the UGV and a remote operator or control centre.
5. Mission-specific payloads: These are specialized tools or equipment that are mounted on the UGV to perform specific tasks. Examples include robotic arms for manipulation tasks or sensors for detecting hazardous materials.
6. Systems integration features such as Graphical User Interfaces (GUI) and remote-control consoles: These features allow operators to interact with the UGV and monitor its performance.

   The functions of these components work together to create an autonomous vehicle that can operate without human intervention. The physical platform provides mobility, while sensors gather information about the environment around it. Navigation, guidance, and control systems use this information to direct movement, while communication links allow for remote operation or monitoring of performance. Mission-specific payloads enable specialized tasks, while systems integration features provide a user-friendly interface for operators to interact with the UGV.

4. List some key societal applications and some scientific applications of UGVs. Why is a UGV is useful for exploration and hazardous missions?

Answer: Some key societal applications of UGVs include transportation, delivery, agriculture, and construction. Autonomous cars and delivery robots are examples of UGVs used in transportation and delivery. In agriculture, UGVs are used as autonomous tractors or harvesters that can work around the clock without human intervention. In construction, UGVs are used for tasks such as excavation or demolition.

Some scientific applications of UGVs include exploration and hazardous missions. UGVs can be used to explore environments that are too dangerous or difficult for humans to access, such as deep sea exploration or space exploration. They can also be used in hazardous missions such as bomb disposal or nuclear clean-up operations where it may be unsafe for humans to work.

UGVs are useful for exploration and hazardous missions because they can operate in environments that are too dangerous or difficult for humans to access. They can also perform tasks that may be too risky for human operators, such as bomb disposal or nuclear clean-up operations. Additionally, UGVs can operate continuously without the need for rest or breaks, making them ideal for long-duration missions.

5. What is the working principle of a UGV?

Answer: The working principle of an Unmanned Ground Vehicle (UGV) involves several components working together to enable the vehicle to operate autonomously. The first component is the physical platform, which includes the chassis, wheels or tracks, and other mechanical components that make up the vehicle. The physical platform provides mobility for the UGV.

The second component is sensors, which are used to gather information about the environment around the UGV. Examples of sensors include cameras, LiDAR radar, and sonar. These sensors provide data about obstacles, terrain, and other features in the environment. The third component is navigation, guidance, and control systems. These systems use information from sensors to determine the vehicle's location (Rykała et al., 2020) and orientation. They also use this information to plan a path for the UGV to follow. The fourth component is communication links. These are used to transmit data between the UGV and a remote operator or control centre. Communication links allow for remote operation or monitoring of performance.

6. Describe with a Block diagram various functional components of an Autonomous Navigation System (ANS) used in UGVs.

Answer:

7. List some popular applications of a UGV for hazardous operations.

Answer: UGVs are commonly used for hazardous operations where it may be unsafe for humans to work. Some popular applications of UGVs for hazardous operations include:

1. Bomb disposal: UGVs equipped with robotic arms and cameras can be used to safely dispose of explosive devices.
2. Nuclear clean-up: UGVs can be used to remotely access and clean up radioactive materials in nuclear facilities.

3. Chemical detection: UGVs equipped with sensors can detect and identify hazardous chemicals in environments that may be unsafe for humans.
4. Search and rescue: UGVs can be used to search for survivors in disaster zones or other hazardous environments.
5. Mine clearance: UGVs equipped with sensors and robotic arms can be used to safely clear landmines or other explosive devices.
6. Firefighting: UGVs equipped with water cannons or other firefighting equipment can be used to fight fires in hazardous environments.

8. What is the difference between UGVs for terrestrial applications and those used for application in space?

Answer: The main difference between UGVs for terrestrial applications and those for space applications is the environment in which they operate. UGVs designed for space applications must be able to operate in the harsh conditions of space, including extreme temperatures, radiation, and a vacuum.

In addition to these environmental differences, UGVs designed for space applications must also be able to navigate and operate in a zero-gravity environment. This requires different types of propulsion systems than those used by terrestrial UGVs. For example, some space UGVs use reaction wheels or thrusters to manoeuvre in zero gravity.

Another key difference is the type of mission that the UGV is designed for. Space UGVs are typically designed for scientific exploration or other specialized missions such as satellite servicing or asteroid mining. These missions require specialized payloads and sensors that are not typically found on terrestrial UGVs.

9. What is off-road trafficability? How is it achieved through sensing and sensor combination?

Answer: Off-road trafficability refers to the ability of a vehicle to navigate through rough terrain, such as mud, sand, or rocky terrain. Achieving good off-road trafficability is important for many UGV applications, such as military operations or exploration. Sensing and sensor combination play an important role in achieving good off-road trafficability. UGVs use a variety of sensors to gather information about the environment around them, including cameras, LiDAR, radar, and sonar. This information is used to create a map of the terrain and identify obstacles that may be in the vehicle's path.

To achieve good off-road trafficability, UGVs must be able to navigate through rough terrain while avoiding obstacles. This requires sophisticated algorithms that can interpret sensor data and plan a safe path for the vehicle to follow. For example, some UGVs use machine learning algorithms to learn from past experiences and improve their ability to navigate through difficult terrain.

Sensor combination is also important for achieving good off-road trafficability. By combining data from multiple sensors, UGVs can create a more accurate map of the environment and better identify obstacles that may be in their path. For example, combining LiDAR data with camera data can help UGVs better identify rocks or other obstacles that may be difficult to detect with just one type of sensor (Matthies et al., 2001)

10. Give a brief classification of the UGV class of vehicles and list the generic criteria for their classifications.

Answer: UGVs can be classified into several different categories based on their size, mobility, and mission requirements. Some common classifications of UGVs include:

1. Micro UGVs: These are small UGVs that are typically less than 10 cm in length and are designed for use in confined spaces or other areas where larger vehicles cannot operate.
2. Small UGVs: These are larger than micro UGVs but still relatively small, typically less than 1 m in length. They are designed for use in urban environments or other areas where mobility is important.
3. Medium UGVs: These are larger than small UGVs and can range from 1 to 3 m in length. They are designed for use in a variety of environments and can be equipped with a range of payloads depending on the mission requirements.
3. Large UGVs: These are the largest class of UGVs and can range from 3 to 6 m or more in length. They are typically used for heavy-duty applications such as mining or construction.

REFERENCES

Appelqvist, P., Knuuttila, J., & Ahtiainen, J. (2010). Mechatronics Design of an Unmanned Ground Vehicle for Military Applications, InTech, 237-261, Mar. 01, 2010. doi: 10.5772/8919.

Bhagwat, G.G., Tandlekar, S.S., Pardhi, N.R. & Dhawale, S.D. (2020). Unmanned Ground Vehicle. IRJET, 7(4).

Gage, D.W. (1995). UGV History 101: A Brief History of Unmanned Ground Vehicle (UGV) Development Efforts. San Diego: Naval Ocean Systems Center. Print.

Gavhane, C.R., & Ansari, A.H. (2013). Unmanned Ground Vehicle. IJREAT, 1(4).

Maheswaran, S., & Murugesan, G. (2020). Unmanned Ground Vehicle for Surveillance. 11th International Conference on Computing, Communication and Networking Technologies (ICCCNT).

Matthies, L., Kelly, A., Litwin, T., & Tharp, G. (2001). Obstacle Detection for Unmanned Ground Vehicles: A Progress Report. Proceedings of the Intelligent Vehicles '95. Symposium, Detroit, MI, USA, 1995, pp. 66-71, DOI: 10.1109/IVS.1995.528259

Niu, X., Yu, T., Tang, J., & Chang, L. (2017). An Online Solution of LiDAR Scan Matching Aided Inertial Navigation System for Indoor Mobile Mapping. Mobile Information Systems, 4802159, 11 pages. https://doi.org/10.1155/2017/4802159

Nowakowski, M., & Kurylo, J. (2023). Usability of Perception Sensors to Determine the Obstacles of Unmanned Ground Vehicles Operating in Off-Road Environments. Applied Sciences, 13, 4892.

Pekka. Appelqvist, J. Knuuttila, & J. Ahtiaine, (2010), 'Mechatronics Design of an Unmanned Ground Vehicle for Military Applications', Mechatronic Systems Applications. InTech, 237–261, Mar. 01, 2010. doi: 10.5772/8919.

Rykała, Ł., Typiak, A., & Typiak, R. (2020). Research on Developing an Outdoor Location System Based on the Ultra-Wideband Technology. Sensors (Basel), 20(21):6171. 10.3390/s20216171. PMID: 33138303; PMCID: PMC7663209.

Ryu, D. et al. (2005). Tele-operation of Field Mobile Manipulator with Wearable Haptic-Based Multi-Modal User Interface and Its Application to Explosive Ordnance Disposal. Journal of Mechanical Science and Technology, 19.

Singer, P. (2009a). Military Robots and the Laws of War. The New Atlantis: A Journal of Technology and Society, 23, 25–45.

Singer, P. (2009b). Wired for War: The Robotics Revolution and Conflict in the 21st Century. New York: Penguin Group.

Velaskar, P., Vargas-Clara, A., Jameel, O., & Redkar, S. (2014). Guided Navigation Control of an Unmanned Ground Vehicle using Global Positioning Systems and Inertial Navigation Systems. International Journal of Electrical and Computer Engineering (IJECE), 4. 10.11591/ijece.v4i3.5183.

Yang, J., Dang, R., Luo, T. & Liu, J. (2015). The Development Status and Trends of Unmanned Ground Vehicle Control System. 2015 IEEE International Conference on Cyber Technology in Automation, Control, and Intelligent Systems (CYBER), Shenyang, China, pp. 1946–1952, doi: 10.1109/CYBER.2015.7288245.

Chapter 11

Crafting a UGV

OVERVIEW

An Unmanned Ground Vehicle (UGV) can be modelled as a system of various subsystems. The prime subsystems that are essential for designing a UGV and their specification are discussed. Various modes of manoeuvre of a UGV, their navigation algorithms, the control software, and its specification are discussed. The functional specification of a UGV and its control system are discussed. To illustrate the manoeuvre commands and controlled usage of sensors, a set of commands are discussed.

11.1 CONCEPTS OF DESIGNING A UGV

Central to designing an Unmanned Ground Vehicle (UGV) is the missions for which it will be deployed and used. Each of the missions can be further broken down into number of workflows or use cases or operating processes. Together, all of these workflows define the concept of a mission. Also, the workflow or operating process will be different for different missions. A mission definition will include at least the following components (Tilbury and Ulsoy, 2010):

Definition and specification of the sensors that will be operational during the mission. The sensors will be automatically turned on when the mission is started and turned off when the mission is stopped or paused. Also, some sensors will be on during some movements and off during other movements.

Definition of processing nodes, their parameters, and their interfaces to other nodes, including topics and services that will be published and subscribed to. It may be noted that certain parameters may be defined at a system level and not necessarily associated to any particular node (for e.g. platform dimensions and speed limits).

Definition of data logs. This will typically include general platform status, specific sensor data streams, diagnostic messages from the various nodes, and mission-specific data such as mode of operation, waypoints, and mission configuration (including changes to such data in the course of the mission). It may be noted that certain system-level logs will need to be maintained irrespective

DOI: 10.1201/9781003432685-11

of a mission actually running, such as change of operational mode (for e.g. from tele operation to manual driving), platform health status, platform configuration changes, and the log of platform pose over time. A mission is a means of recording the details of the specific operational state of the platform over a specific duration for offline analysis.

Change or reconfiguration of mission configuration will only be allowed for the Administrators and Developers user groups. The Operators user group may run preconfigured missions. Following are some distinct application workflows to be implemented for different applications:

(i) Autonomous navigation: In this workflow, the UGV is able to navigate autonomously through a predefined operational area using sensors and algorithms to detect and avoid obstacles. This workflow could be used for applications such as surveillance, mapping, or inspection (Kandath et al., 2018).
(ii) Payload delivery: In this workflow, the UGV is equipped with a payload bay or attachment mechanism to carry and deliver payloads to specific locations within the operational area. This workflow could be used for applications such as logistics, delivery, or material handling.
(iii) Remote operation: In this workflow, the UGV is remotely operated by a human operator using a control station. The operator is able to control the movement and actions of the UGV in real time. This workflow could be used for applications such as search and rescue, hazardous material handling, or exploration (Fabris et al., 2021).
(iv) Data collection: In this workflow, the UGV is equipped with sensors and instruments to collect data from the operational area. The data collected could include environmental measurements, images or video, or other types of data relevant to the application. This workflow could be used for applications such as environmental monitoring, surveying, or scientific research (Nowakowski et al., 2023).

11.2 MANUAL DRIVING MODE

In this workflow, the platform will be manually driven after disengaging the Drive-by-Wire (DBW) system. The disengagement of the DBW system will be tool-less and will not require disassembly. The disengagement of the DBW system be detected in software. and actuation of the platform controls will be disabled.

A mission (as described above) may be configured in this workflow. A mission may be paused, resumed, stopped, and started multiple times in the course of a single operation of the workflow. This workflow will typically be used for the following purposes:

a. Acquire data sets for offline evaluation of certain processing algorithms.
b. Evaluate the online performance of certain processing algorithms.
c. Create an initial occupancy map of the operational area to be used in subsequent autonomous runs.
d. Acquire waypoints for subsequent autonomous runs.
e. Retrieve the platform in the course of an autonomous mission where the autonomous navigation system is unable to move the platform; the autonomous mission will be paused, the platform will manually be brought to a nominal state, and the autonomous mission resumed.

The GUI (Graphical User Interface) will display a geo-referenced image of the operational area. The pose of the platform, if available, will be displayed as an overlay on this image. The platform health status, if available, will be displayed suitably. Diagnostic messages will also be displayed appropriately.

In manual driving mode, it is important to ensure that the operator has a clear and unobstructed view of the operational area. This can be achieved through the use of cameras and other sensors mounted on the UGV, which provide real-time visual feedback to the operator.

The operator should also have access to real-time data on the UGV's performance and status, such as battery level, speed, and sensor readings. This information can be displayed on the GUI to help the operator make informed decisions (Shuzhi, 2006) while driving the UGV (Parasuraman et al., 2017).

Manual driving and mission initiation may be performed by any member of the Operators user group. Mission configuration, however, can be carried out only by a member of the Developers or Administrators user group.

11.3 TELEOPERATION

In this workflow, the DBW system will be engaged, and the platform will be driven from the Mission Control System (MCS). The range of teleoperation should cover the predefined area of operation, with the MCS placed at an arbitrary location in the area of operation.

The GUI will display a geo-referenced image of the operational area. The pose of the platform, if available, will be displayed as an overlay on this image. The platform health status, if available, will be displayed suitably. The sensor feed and camera feeds of designated sensor mounted on the platform will be automatically visualized at the MCS.

a. If the platform is being driven forward; the front centre camera (Cl) feed should be displayed
   i. If the platform is moving forward and turning to the right, the side-right camera (C4) feed should also be displayed

ii. If the platform is moving forward and turning to the left, the side-left camera feed (C5) should also be displayed

b. If the platform is moving in reverse, the front (C1) and reverse camera (C6) feed should be displayed.
c. The user should be able to override the automatic selection to show either the side-right (C4) or side-left (C5) camera feed on a need basis. In conjunction with the front centre camera feed, the user should have the provision to display the feeds from the thermal cameras (Tl & T2) instead of the colour cameras, for teleoperation at night. The user should also have the provision to display the feeds from the Cl and Tl or C2 and T2 cameras.
d. A mission may be configured in this workflow to acquire, process, visualize, and log data from selected sensors. A mission may be paused, resumed, stopped, and started multiple times in the course of a single operation of the workflow.
e. Teleoperation and mission initiation may be performed by any member of the operator's user group. Mission configuration, however, can be carried out only by a member of the Developers or Administrators user group
f. Autonomous navigation to a target location
g. In this workflow, the user will designate a target pose for the platform to achieve (consisting of a desired position and orientation of the platform), either through the MCS or the on-board DI. This location may be either a stored location or a fresh location. If the location is marked afresh, the user will have an option to label and save it for later reuse. The range of operation for this and the subsequent autonomous workflows will be the same as for teleoperation.

A mission will be configured in this workflow to acquire and process data from selected sensors to build an environment map (termed online map). The mission may also take as input a map created from an earlier run of the platform (termed offline map): The online and offline maps will be used together with the user-defined target location to create a path for the platform to follow. The path plan will be used to generate motion commands for the platform closed-loop control system to follow. The mission configuration may also include logging of specified data collected during the execution of the mission, including but not limited to, the pose of the platform, selected sensor and processed data, log messages, and health status of the platform.

A mission may be paused, resumed, stopped, and started multiple times in the course of a single operation of the workflow. The target location may be modified by the user when the mission is stopped or paused. The mission may be reconfigured when the mission is stopped. Mission parameters may be reconfigured at any time.

The user may switch to teleoperation or manual operation at any time in the course of the mission, without necessarily pausing or stopping the mission (for e.g. in an emergency). Suitable mechanisms should be incorporated into the workflow to cater for this eventuality.

The GUI will display a geo-referenced image of the operational area. The pose of the platform will be displayed as an overlay on this image. The platform health status will be displayed suitably. The target location will be input through the GUI and displayed appropriately. The path planned for the platform will also be displayed. The environment map created for navigation will be displayed on demand, as an overlay to the image. The GUI will also provide a display of features of interest, such as dynamic obstacles, pedestrians, and so forth. The user may also choose to display the camera feed from up to two cameras on the platform. The user may choose to display the cost-map generated by the platform.

Mission initiation and waypoint specification may be performed by any member of the Operators user group. Mission configuration, however, can be carried out only by a member of the Developers or Administrators user group.

11.4 AUTONOMOUS NAVIGATION OVER A NON-CYCLIC ROUTE

This workflow will be similar to the workflow as in teleoperation mode above, with the exception that the user will input a sequence of waypoints through which the platform will navigate autonomously. This sequence of waypoints may be either a stored sequence or a fresh sequence. Waypoints may be marked on the GUI itself or loaded through a file in the Route Data Definition format or Route Navigation data format. If the sequence is marked or loaded afresh, the user will have an option to label and save it for later reuse.

The waypoints will be pre-processed and fed to the autonomous navigation system, according to the following rules:

a. If the platform is already some distance along the defined route when the mission is started or resumed, the waypoints “behind” the current position of the platform will be discarded and the remaining points “ahead” of the platform will be fed in sequence.
b. If any of the waypoints lies on an obstacle, the waypoint will be displaced along the defined width of the path until it is clear of the obstacle, with consideration to the platform width.
c. If it is not possible to displace the waypoint suitably, that waypoint will be discarded and the next waypoint will be fed to the system.

d. If the distance between two consecutive waypoints is greater than a certain threshold, additional intermediate waypoints could be automatically generated to ensure smooth and safe navigation of the UGV.
e. If the UGV is unable to reach a waypoint due to obstacles, the system could automatically generate a different route to reach the next waypoint.

11.5 AUTONOMOUS NAVIGATION OVER A CYCLIC ROUTE

This workflow will be similar to the workflow as described in the teleoperation mode Section 11.3. Preprocessing and feeding of the waypoints will be similar to as outlined in the workflow in Section 11.4. The difference will be in the feeding of the waypoints to the autonomous navigation system, where the last waypoint in the route will be followed by the entire sequence again from the beginning, in a continuous loop, until the mission is stopped.

In a cyclic route, the UGV will repeat the same sequence of actions and movements over and over again, which can be useful for applications such as patrolling or monitoring. In a linear route, the UGV will follow a specific path to reach its destination, which can be useful for applications such as delivery or transportation (Rubio et al., 2019) between fixed source and destination.

11.6 MISSION REPLAY AND ANALYSIS

This workflow will be purely offline and will involve the replay of the logs recorded from a previous mission run in any of the workflows above. The GUI of a UGV remote station must provide options for the visualization of the mission execution as if the platform were currently executing the mission. Provision must be available to start, stop, fast-forward, rewind, and move to a specified time in the mission. Visualization will include at least the pose of the platform, the health status of the platform, camera feeds, cost map, path plan, and diagnostic messages. A mission can be replayed by any authenticated user.

11.7 SPECIFICATION OF BASE PLATFORM

The chosen base platform should be manually drivable, Ackerman-steered, four wheeled, and battery operated with suspension on all wheels.

The base platform should be designed with modularity and flexibility in mind. This allows for easy customization and adaptation to different missions and environments. Modular design also facilitates maintenance and repair, as individual components can be easily replaced or upgraded.

The desired performance specifications of the platform are given below. It may be noted that the performance specifications stated below are for the

Table 11.1 Specifications of base platform with all components mounted

Sl. No.	*Description*	*Specification*
1.	Minimum turning radius	Less than 5 meters
2.	Maximum speed	At least 40 kilometres per hour
3.	Gradeability	At least 15% fully laden
4.	Ground clearance	At least 0.15 metres
5.	Passenger seating capacity	At least 2 persons (200 kgs) after all modifications required under this contract
6.	Additional payload carrying capacity	At least 50 kgs excluding passengers and equipment mounted as a part of this contract
7.	Braking	Parking brake or similar mechanism for holding platform position on power off
8.	Gear shift	One forward and one reverse
9.	Endurance on single charge	At least 80 kilometres on flat terrain

fully integrated platform with all components, defined in subsequent sections, mounted, and operational (Table 11.1).

The reference platform for this contract is the Mahindra E20.

The platform will need to be suitably modified to accommodate the components defined in subsequent sections.

11.8 DRIVE-BY-WIRE

The chosen platform will need to be modified for computer-controlled actuation of the platform driving controls, for example, steering, accelerator, brake, and gear shift. In addition, the ancillary controls, such as platform power on, headlights, turning lights, and horn, should also be computer controllable. The DBW system needs to meet the following requirements.

1. Electronic/Electromechanical Actuation (EMA) of platform driving controls and ancillary elements. EMAs should include current limiting to prevent power drive burnout or shutdown. EMAs should be provided with limit sensing for the allowable travel of the EMAs. The performance of the driving controls in DBW mode should be equivalent to the performance in manual driving. For example, providing a full brake command in DBW mode should bring the vehicle to a stop within the braking distance when the platform is driven manually (Guidi et al., 2021).

2. Conversion between manual and DBW modes should be tool-less and should not require any disassembly. Disengaging of the DBW system should be detectable in software.
3. Provision of absolute position feedback for driving control elements in addition to position feedback of actuators.
4. Provision of a magnetic wheel speed sensor on one of the wheels needs to be incorporated.
5. Provision of a ring of ultrasonic sensors around the platform.
6. Provision of a light-sensitive sensor, to determine the ambient light conditions for automatic tum-on and turn-off of the platform headlights.
7. Development of embedded software for the following functions on dedicated embedded hardware:
   a. Power on self-calibration of the driving control element positions.
   b. Closed-loop control of driving control elements with safety interlocks for maximum allowable range of motion and current limits.
   c. Open-loop control of platform speed and heading rate.
   d. Closed-loop control of platform speed and heading rate with safety interlocks for maximum speed and corresponding yaw rate. Feedback of platform speed and heading will be taken from the localization module.
   e. Safety interlock between absolute position feedback and position derived from EMA feedback to detect and handle EMA failures.
   f. Safety interlock between the inputs from the ultrasonic sensors and the platform driving commands to ensure low-latency, reactive, obstacle detection and avoidance. Provision should be made to override this interlock through a software configuration switch.
   g. Operational interlocks between accelerator, brake, and gear shift.
   h. Operational interlocks to prevent EMA and ancillary controls operation when in manual drive mode.
   i. Hardware emergency stop (e-stop) (see point in the chapter) shutdown sequence should bring the platform to a rapid halt within automotive braking safety standards.
   j. Software e-stop or any other failure condition should bring the platform to a smooth halt within ergonomic standards.
   k. Automatic turn-on and turn-off of platform headlights based on ambient light conditions. The threshold at which the lights a~e turn on and off should be configurable.
   l. Automatic turn-on and turn-off of platform turning indicators, based on the temporal variation of platform motion commands (speed and yaw rate) from a higher-level controller.
   m. ROS-based software interfaces for input of driving and ancillary control commands.

n. ROS-based software interface for input of command platform speed and yaw rate.
o. ROS-based software interface for input of software e-stop commands.
p. ROS-based software interface for providing feedback on current actuator positions, driving control elements positions, ancillary controls status, DBW system status (including e-stop status), and ultrasonic sensor readings, both in manual drive mode and DBW mode.
q. ROS diagnostics-based interface and instrumentation for collecting DBW system performance data, including hardware status, fault conditions, closed-loop control rates, and command-to-actuation
r. ROS parameter and dynamic reconfigure interfaces for run-time specification and modification of parameters, including closed-loop rates, nominal actuator, and driving control element home positions and limits, closed-loop gains, and maximum platform speed and yaw rate.
s. Development of a safety-critical software architecture that conforms to ISO 26262 safety standards.
t. Implementation of a fault-tolerant system design that can detect and recover from software and hardware failures.
u. Development and integration of a sensor fusion algorithm to combine data from multiple sensors, including ultrasonic sensors, Light Detection and Ranging (LiDAR), and cameras, to improve obstacle detection and avoidance.
v. Implementation of a fail-safe mechanism to prevent unintended acceleration or deceleration of the platform.
w. Development of a comprehensive testing and validation plan to ensure the safety and reliability of the DBW system under various operating conditions and failure scenarios.
x. Integration of user interface for displaying critical system information, such as current speed and heading, status of safety interlocks, and fault conditions.
y. Implementation of a logging and data analysis system to record system performance data and facilitate post-mortem analysis in case of system failures.
z. Integration of a remote monitoring and control system to enable remote operation and monitoring of the DBW system.

8. Wireless e-stop with a dedicated, long-range (> 1 km NLOS), communication link and physical e-stop switches mounted on the platform dashboard and outside. Triggering any of the e-stops should automatically bring the platform to a rapid halt without software intervention. A signal should simultaneously be provided to the embedded computer software, which should instantly release the accelerator, depress the brake, and

power off the platform. On release of the e-stop, the actuators should be brought back to their nominal positions, without powering on the platform. The wireless e-stop communication link should be continuously monitored. The platform DBW should be disabled if the wireless e-stop communication link breaks.

11.9 SPECIFICATION OF SENSORS REQUIRED FOR A MISSION

The following classes of sensors will need to be procured and mounted on the platform for environment perception. It may be noted that the reference models for the sensors are specified for performance requirements, and it is expected to choose equivalent sensors that also meet the necessary environmental requirements for passing the tests specified in Section 11.14. However, if no sensor is commercially available that meets the performance and environmental ·requirements, the sensor that best meets the performance and environmental specifications will be chosen. This sensor should be subjected to environmental tests for use in the UGV.

The scope of UGV design includes the fabrication and integration of suitable mounts for the sensors on the platform, definition of coordinate frames for each sensor and the platform, and calibration of the coordinate frame transformations from each sensor to the platform frame. Additionally, the sensors will themselves need to tested and evaluated for performance parameters such as range and resolution, and if necessary, calibrated. Stereo camera mounts must be fabricated to ensure accurate alignment. The sensor mounts and cable harnesses should be designed to facilitate easy removal and remounting of sensors (Apestigue et al., 2022).

The mounting pose (position and orientation) of all the other sensors should be calibrated relative to the IMU centre of rotation. Further, the IMU centre of rotation should be calibrated relative to the platform dimensions. The positions of the phase centres of the GNSS antennas relative to the IMU centre of rotation should be calibrated (Kozlov et al., 2023).

The rationale behind the mounting is to minimize dead zones around the platform and so that each zone around the platform is observed by at least two sensors. It is expected to carry out a thorough analysis of the placement of sensors with respect to these criteria as shown in Table 11.2 (Majid et al., 2022).

11.10 EMBEDDED COMPUTERS AND INTERFACE HARDWARE

At least six homogeneous embedded computers, with at least the following specifications, will need to be provided on the platform for sensor data acquisition and processing.

Table 11.2 Sensor specifications of a UGV

Sl No	*Sensor*	*Reference Mode*	*Requirement*
1	3DLidar	Velodyne VLP-16	One sensor each mounted on either end of the front edge of the roof and one sensor mounted centrally on the back edge of the roof (total 3 nos.), tilted downwards by 10°
2	Colour camera	PointGreyBlackFly BFLY-PGE-23S6M-C	Three cameras (C1, C2, C3) mounted in a triangular configuration in the vertical plane facing forward; one camera placed centrally at the front upper edge of the windscreen (C1), and the other two (C2 and C3) placed approximately at either end of the bottom edge of the windscreen, with an adjustable baseline; synchronized capture from all three cameras through hardware trigger; cameras to be calibrated for stereo vision; 8 mm and 12.5 mm lenses for cameras; all cameras to have global shuttering
			One camera each on either side of the platform (C~ and C5), towards the rear facing forward and down; 12.5 mm lenses for cameras
			One camera mounted centrally facing the rear (C6); 8 mm lens for the camera
			C4, C5, and C6 will be used primarily for teleoperation and are required to have a minimum resolution of 640 x 480 pixels, with a minimum frame rate of 24 fps. These cameras may be chosen to be different from C1, C2, and C3.

(*Continued*)

Table 11.2 (Continued)

Sl No	*Sensor*	*Reference Mode*	*Requirement*
3	Thermal camera	XenicsServal 640 GigE	One camera each mounted centrally facing forward (T1) and rear (T2); minimum resolution of 640 x 480 pixels and minimum frame rate of 24 fps, 5.8 mm lens; these cameras will be used for teleoperation, especially in night-time conditions
4	Time-of-flight camera	Fotonic E70	One camera each facing forward and rear
5	RADARs	UMRR-0C with type 128 antenna (at 77 GHz)	Multiple sensors mounted to provide a 120o horizontal field-of-view, both in front of and to the rear of the platform; range of at least 100 m ahead of the platform and 50 m behind the platform. Sensor to be capable of resolving in range, velocity, and angular separation between objects
6	Ultrasound sensors	Maxbotix Max sonar E20XL	Multiple sensors mounted in a ring around the platform to ensure no dead zone beyond ~1 m from the platform; max range of at least 7 m Sensors to be coupled directly to the DBW subsystem for low-latency, reactive obstacle avoidance; software configuration switch for runtime override of the coupling.

7	Integrated GNSS (INS, GNSS antennae, wheel speed sensor, RTK base station, and communication infrastructure	Advanced Navigation Spatial Dual	GNSS receiver and IMU mounted inside the platform; IMU centre of rotation centrally located above the rear axle; receiver configured for dual antenna static alignment, GAGAN SBAS and online RTK. Two GNSS antennae mounted on roof as per recommended spacing, and to be integrated with the spatial system; antennae to support at least GPS/GLONASS constellations (additionally IRNSS if available at the time of procurement.) One wheel speed sensor to be mounted on one of the non-driven wheels; output to be interfaced to the spatial system; reference model iMAR-iMWS v2 Base station and with the communication infrastructure components for RTK to be set up and integrated with both receivers; regulatory approvals, if any, for the same to be obtained from appropriate authorities. Online RTK solution to be established and integrated with the spatial system Navigation solution must be established through static alignment

a. Intel i7 processor or better with at least 8 GB on-board soldered RAM.
b. Dual Gigabit Ethernet ports.
c. At least 4 USB 2.0 or higher ports
d. At least 128 GB SSD.

The computers should be networked over a Gigabit Ethernet backbone, through a managed switch. Both Gigabit Ethernet ports from each computer will be connected to the switch. All Ethernet-based sensors will also be connected to the switch. The switch should permit configuration such that network-based sensor data are routed to specific computers without affecting the throughput between the computers. Four additional ports should be provided on the switch as spares.

The computers should be housed in a suitable enclosure with cables from individual sensors terminating on connectors on the enclosure. Suitable cable harnessing and routing will need to be provided inside the enclosure for power and data interfaces to designated computers. The spare Ethernet ports from the switch should be available on connectors terminated on the enclosure. Further cabling should be provided to route the spare Ethernet ports close to the dashboard at the co-driver's seat. One USB port from each computer should be available jointly on a single connector terminated on the enclosure. Breakout cables for each USB port should be provided on the outside. Suitable provision must be made to interface the Platform-Mounted Display (PMD) to all the embedded computers, through an embedded KVM switch and connectors available on the enclosure. Indication must be available on the enclosure to determine the embedded computer currently connected to the display. The enclosure and connectors should be easily accessible.

11.11 WIRELESS COMMUNICATION INTERFACE

A wireless communication interface should be provided between the platform and the Master Control Station (MCS). The range and bandwidth of the link should meet the following requirements:

a. Range: The link should provide unbroken connectivity between the platform and the MCS within the area of operation, irrespective of the relative locations of the platforms and the MCS. Broadly, the link must have a range greater than 500 m non-line-of-sight, across buildings for operating in urban terrain.
b. Bandwidth: The link should provide sufficient bandwidth for exchange of the following data at all times
    a. At least two camera feeds from the platform to the MCS, with a resolution of at least 640 x 480 pixels @ at least 10 frames per second. The

camera feeds may be compressed at the platform end before transmission. Suitable open-source codes should be used for compression.

b. Platform health status from the platform to the MCS -1 KB @ 10 Hz
c. Platform control commands from the MCS to the platform -l KB @ 20 Hz

3. In addition to the above, on-demand visualization of the following data will be required at the MCS.

a. Occupancy map generated on the platform 500 x 500 grid @ 2 Hz
b. Diagnostic, debug, and performance feedback messages @ 2 Hz

The implementation can consider the use of multiple data links to meet these requirements, or intelligent data. Traffic management to ensure high-priority data is always communicated, while low-priority data is scheduled for exchange based on bandwidth availability. The link should provide encryption equivalent to at least WPA2 for Wi-Fi links. The interface to the communication link from the platform and MCS-embedded system should be ETH-based.

11.12 EMBEDDED SOFTWARE

Embedded software development will be carried out using the ROS Kinetic framework on the Ubuntu 16.04 Linux distribution. The software development effort will include the following tasks:

1. Time synchronization on all embedded computers and with the MCS
2. Development of interface and data acquisition nodes for all sensors; all configurable sensor parameters should be exposed as static parameters and through dynamic reconfigure interfaces. Data acquisition nodes should preprocess the data to perform necessary filtering of data, such as removal of out-of-range readings and restriction of the data to a configurable region of interest. Data from LiDAR sensors should be transformed to point clouds in the world frame. Data from stereo sensors should be preprocessed to generate disparity images and point clouds in the world frame. Preprocessed data should be made available on topics using standard or, if necessary, custom ROS messages. Raw data from all sensors should be available for logging through ROS bag files (Chen et al., 2020).
3. Identification and incorporation of diagnostics for runtime and offline analysis of system performance, including data interfaces (software and hardware) and data processing algorithms; this will include mechanisms of determining, logging, and analysing the processor and memory usage of the various components, publishing/subscribing/processing rates with detection and logging of dropped messages

4. Implementation of the workflow for teleoperation, including parameterization, runtime (re)configuration, and automatic launching and monitoring of processes
5. Implementation of the UGV workflow for autonomous operation, including parameterization, runtime (re)configuration, and automatic launching and monitoring of processes; here it is to be noted that the development is required to do the following tasks
   a. Generation of platform pose and speed estimates (localization module)
   b. Generation of occupancy maps from sensor data (perception module)
   c. Platform path planning and generation of platform motion commands (navigation module). Working code for the three modules defined in a, b, and c above has been developed indigenously This code will need to be integrated with the workflow, possibly with the addition of required data, logging and diagnostic interfaces, and incorporation of error handling. The code for these three modules will be integrated with the main control software of the UGV.
6. Implementation of the workflow for offline playback of logged data; this will include aggregation of the logs for a particular mission and automatic generation of necessary launch files to run ancillary processes (for e.g. nodes to process raw sensor data)
7. Identification of failure modes and incorporation of suitable interlocks, workarounds, fallbacks, and fail-safe actions to deal with failure

11.13 USER INTERFACE

The user interface will comprise two distinct components: a Platform-Mounted Display (PMD) and an MCS. Both will feature a common Graphical User Interface (GUI).

11.13.1 Platform-mounted display

The PMD will consist of a sunlight-readable, touch-enabled display (at least 10"). This will be connected to the platform embedded computers through the KVM switch. The display positioning and mounting should be ergonomic.

11.13.2 Master control station

The MCS is required to be mobile to the extent that it is portable from one location to another rapidly. During operation of the system, the MCS will remain static. It will feature at least one sunlight-readable, touch-enabled display, together with a keyboard, mouse, and an embedded computer. It will also feature a console emulating the platform manual driving interface, with at least a steering wheel, accelerator, and brake pedals, together with seating for one operator. The positioning of the various components should be ergonomic. The operation of the steering wheel and pedals should closely correspond to that on the actual platform. The MCS should have an on-board rechargeable battery-based power pack for at least four hours of endurance with all components of the MCS operational. The charger for the power pack should be on board and should permit online charging through a standard 220 volt wall socket, while the loads are operational. The charger should be capable of optimizing the charging cycle of the attached battery as per the battery type. One set of spare batteries with charger and battery monitoring system needs to be kept as a spare for any UGV operation. The packaging for the batteries must meet extant regulations governing the packaging for transport of such equipment, especially batteries (Liang et al., 2019).

The MCS will communicate with the platform through a wireless communication link.

11.13.3 Graphical user interface and application software

The GUI should be cross-platform (preferably web-based; for Windows 7 and later, Ubuntu Linux 14.04 and later, and Mac OSX 10.0 and later), and follow a three-tier model-view-controller architecture. The GUI and application software will have at least the following features:

1. Facility for loading and displaying a geo-referenced image of the operational area with panning and zooming facility, and an overlay of the current platform pose on the image
2. Display of an environment occupancy map with panning and zooming facility and an overlay of the current platform pose in the map
3. Switching between tele-operation and autonomous modes
4. Selectable display of up to two camera feeds from the platform
5. Two tele-operation interfaces: (1) Keyboard based and (2) Hardware console based; the hardware-based interface will only be available at the MCS
6. Definition, storage, and retrieval of a route map for the platform in terms of a sequence of connected waypoints; the route map definition should support RDDF, RNDF, and plain text formats
7. Definition, storage, and retrieval of landmarks in the map in terms of specific locations
8. Definition, storage, and retrieval of a mission in terms of the sensors to be used, processing nodes to be initiated on specific computers on the platform, parameter configurations for the processing nodes, the route map to be used, and logging requirements

9. Storage, retrieval, and editing of an environment occupancy map; providing the occupancy map as an initialization to the application workflows
10. Facility for starting, pausing, resuming, and stopping a mission; configuration parameters and process launching should be propagated appropriately through the system to bring up the necessary processing nodes with the relevant configurations
11. Filtered display and logging of platform health status and diagnostics
12. Facility of replaying a logged mission
13. Automatic session management and interlocks with multiple instances of the user interface running in parallel
14. User creation, deletion, and management with profiles for administration of the UGV need to be developed with password-based user authentication
15. Automatic detection of manual mode and appropriate disabling of user interface elements need to be a feature of the UGV
16. Automatic detection of low-power conditions (platform and power pack), of aborting the current mission, and of returning to a designated location for recharging

The GUI should adapt to variable display sizes and/or dual-head configurations, with dockable and free-floating options for various panes, and should be user friendly.

11.13.4 Power supply, distribution, and management

A rechargeable battery-based power pack will need to be provided for on-board electronics including the DBW system, sensors, embedded computers, and platform-mounted display. The power pack should provide for at least four hours of endurance on full load with all systems operational, including the platform DBW system (Boukoberine et al., 2019). The charger for the power pack should be on board and should permit online charging through a standard 220 volt wall socket, while the loads are operational. The charger should be capable of optimizing the charging cycle of the attached battery as per the battery type. The power pack should provide a single DC output voltage, which should be used to generate appropriate voltages required for various subsystems through suitable electronics (for e.g.: through DC to DC converters). Additionally, the power pack should provide an AC 220 volt output through an inverter, sufficient to power two laptops simultaneously, The power points for these outputs should be available close to the dashboard at the co-driver's seat. Individual fuses and software-controllable switches should be provided to control power input to each individual component (DBW, embedded computers, sensors etc.).

The power pack should provide a software interface for switching on individual loads, reading the capacity of the current rate of discharge. The software interface should be monitored on an embedded computer. The battery pack should be hot-swappable and mounted suitably to permit replacement of depleted batteries with fully charged batteries without powering off the system hardware and software interfaces of the power pack.

11.14 PLATFORM RUGGEDIZATION

The integrated platform is required to operate in all weather, outdoor, unpaved off-road, and paved on-road conditions. The MCS is required to operate in all-weather outdoor conditions. The mechanical hardware, including sensor mounts and enclosures, should be appropriately surface treated and powder coated to remain rustproof.

In addition to ruggedization of the hardware, software and electrical systems must also be designed and implemented with the same level of robustness. This includes implementing appropriate error handling and fault-tolerance mechanisms to ensure that the system can continue to operate effectively in the event of unexpected failures or malfunctions.

The integrated platform and MCS are required to pass the following qualification tests. Performance, electrical, and mechanical checks will be carried out on the integrated platform as defined in the relevant test procedures. It may be noted that the tests are meant for qualification and not for certification, and will be carried on the integrated platform and MCS as a whole, with all components mounted and functional, and not for each individual component. The tests should be carried out at an agency certified and approved to carry out the relevant tests.

1. High temperature: JSS 55555 Test No. 17, Procedure 5, Condition G (+ 55° C ± 3° C for 16 hours)
2. Low temperature: JSS 55555 Test No. 20, Procedure 4, Condition M (00 C for 16 hours)
3. Damp heat: JSS 55555 Test No. 10 (40° C ± 30° C, 95% RH for 16 hours)
4. Dust: JSS 55555 Test No. 14 (1 hour)
5. Driving rain: JSS55555 Test No. 12 Condition C (450 and horizontal water jets for 1 hour)
6. Roadability:
   a. JSS5555 Test 29, Procedure 1 (platform and components in packaged condition mounted on a carrier vehicle)
   b. JSS5555 Test 29, Procedure 1 (integrated platform only)
7. Radiated emissions: MIL-461F RE 102 (up to 18 GHz)

11.15 MISSION STATEMENT OF A UGV

Top view of a UGV fitted with microwave proximity sensors in the periphery for collision avoidance, front and rear camera for visual Simultaneous Localization and Mapping (SLAM), left & right camera for omni-directional movement, top LiDAR scanner for 360° scanning of the environment for 3D point cloud generation and Digital Surface Model (DSM) generation. The UGV bottom is fitted with probes and scanners for collection of Nuclear, Biological and Chemical (NBC) concentration data from the contaminated zone. Refer to Figure 11.1 for the top view of the UGV.

1–16 proximity (microwave) proximity sensors for collision avoidance
TC-thermal camera for scanning of the environment in the absence of daylight
LiS- LIDAR Sensor
MWPS – Microwave proximity sensor
FC_1, FC_2 -> Front camera
RC_1, RC_2 ->Rear camera

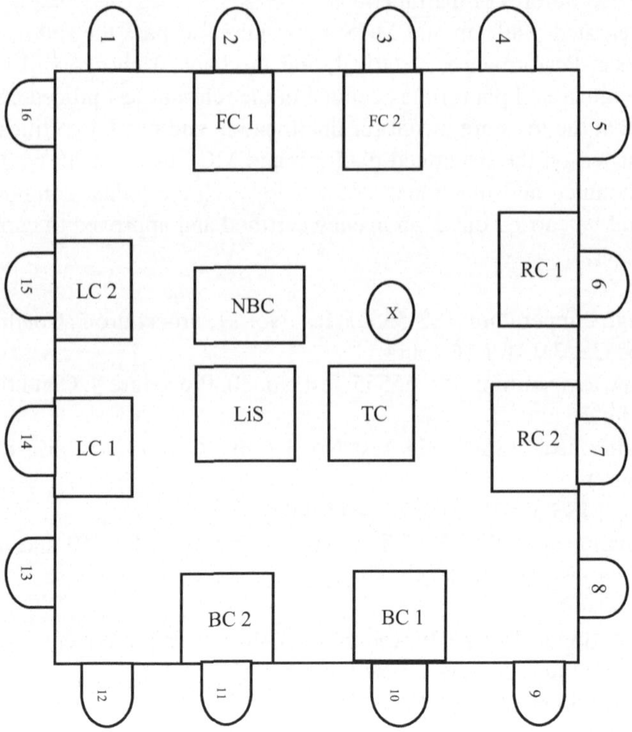

Figure 11.1 Top view of a UGV for NBC recci and collection mission.

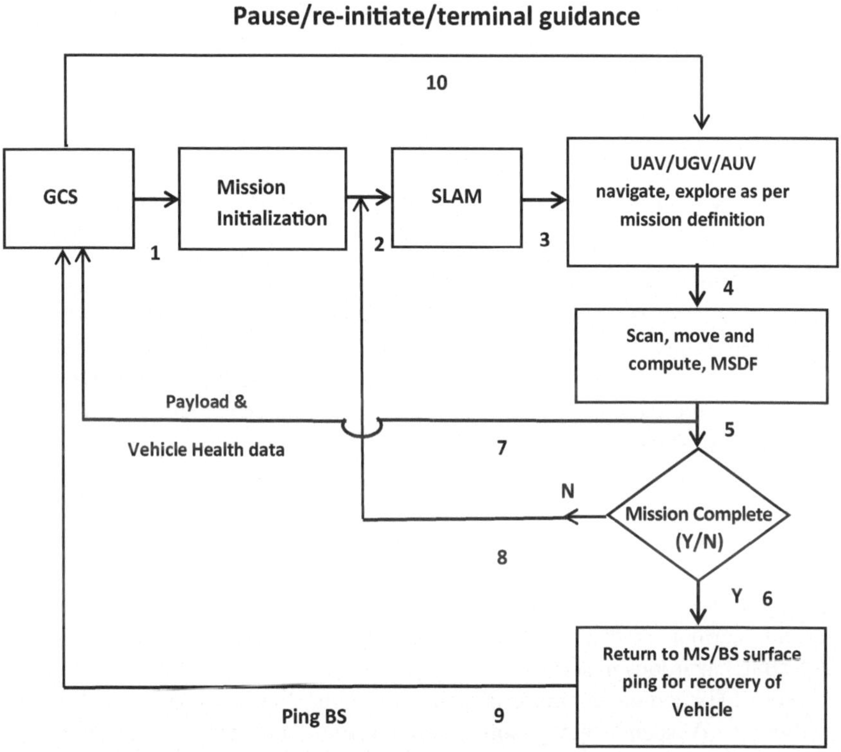

Figure 11.2 Control flow diagram depicting the UGV.

The sensors are placed and positioned on the body geometry of the UGV for uniform sampling of the environment. Not all the sensors are active and sensing during the mission. The sensors are activated and deactivated depending on mission operation and outcome of motion and SLAM. The complete UGV mission can be described through the following pseudo-code with reference to the control diagram depicted in Figure 11.2.

1. If (forward move), then open (FC_1, FC_2, LiS, 1, 2, 3, 4); other sensors are off.
2. If (forward & right move), then open (FC_2, FC_1, 2, 3, 4, 5, 6, LiS)
3. If (forward & left move), then open (FC_1, LC_1, 2, 1, 16, 15, LiS)
4. If (move-back) then open (BC_1, BC_2, 9, 10, 11, 12, LiS)
5. If move (back & left)
6. If (night operation), then open scan (TC, LiS)
7. If (ground sample collection mode is on), then stop the vehicle and stop all sensors except the NBC sample collectors mounted in the bottom of the UGV to collect NBC data.

Forward motion if (forward move), then on (1, 2, 3) & on (FC_1&FC_2) & scan LiS on (7, 8, 9), & if (turn right), then on (FC_1, FC_2) & (RC1 & RC2) & if (turn left), then on (9, 10, 11, 12) & (FC_1& ...)

QUESTIONS AND ANSWERS

1. What are the key design concepts of a UGV? What are different components of a UGV?

Answer: The key design concepts of a UGV are centred on the mission for which it will be deployed. Each mission can be broken down into workflows or use cases that define the concept of a mission. The operating models of a UGV can vary depending on the mission requirements. By understanding the workflows and operating processes of a mission, you can create a UGV that is efficient and effective.

The different components of a UGV include sensors, embedded computers, software, and user interfaces. These components work together to enable the UGV to perform its intended mission. Sensors are an essential component of any UGV as they provide data about the environment in which the vehicle is operating. Examples of sensors used in UGVs include LiDAR sensors for mapping terrain, thermal cameras for detecting heat signatures, and chemical sensors for detecting hazardous materials.

Embedded computers are responsible for controlling the various subsystems within a UGV such as navigation, communication, and sensor data processing. These computers should have high processing power and memory capacity to handle complex tasks such as autonomous navigation. Software is another critical component of a UGV as it provides the instructions necessary to control the vehicle's subsystems. Embedded software used in a UGV includes tasks such as sensor data processing, path planning algorithms for autonomous navigation, and communication protocols with remote operators.

User interfaces allow human operators to interact with the UGV and provide input commands or receive feedback from various subsystems. Different user interface components of a UGV may include displays for sensor data visualization, joystick controllers for manual operation, and touchscreens for system configuration. By understanding these different components and designing them to work together effectively based on specific mission requirements, you can create an efficient and effective UGV that meets your needs.

2. What are different operating models of a UGV?

Answer: There are several different operating models of UGVs, each with its own advantages and disadvantages. The choice of operating model will depend on the specific mission requirements and the capabilities of the UGV. One common operating model is teleoperation, where the UGV is controlled by a human operator using a remote control or joystick. This allows for precise control over the vehicle's movements and actions, but requires a skilled operator and can be limited by communication range and bandwidth.

Another operating model is autonomous operation, where the UGV uses sensors and algorithms to navigate and perform tasks without human intervention. This can be useful in situations where it is not safe or practical for a human operator to be present, such as in hazardous environments or during long-duration missions. However, autonomous operation requires sophisticated algorithms and sensors to ensure safe and effective operation. A hybrid operating model that combines teleoperation and autonomous operation may also be used in some situations. For example, an operator may use teleoperation to guide the UGV through a complex environment or obstacle course, while allowing the vehicle to operate autonomously during simpler tasks such as patrolling or surveillance.

3. Give a brief technical specification of various subsystems used in a UGV. Describe the functions of these subsystems.

Answer: There are several subsystems used in a UGV, each with its own technical specifications and functions. Some of the key subsystems include:

1. Sensors: Sensors are used to provide data about the environment in which the UGV is operating. Examples of sensors used in UGVs include LiDAR sensors for mapping terrain, thermal cameras for detecting heat signatures, and chemical sensors for detecting hazardous materials.
2. Embedded computers: Embedded computers are responsible for controlling the various subsystems within a UGV such as navigation, communication, and sensor data processing. These computers should have high processing power and memory capacity to handle complex tasks such as autonomous navigation.
3. Power systems: Power systems are responsible for providing energy to the various subsystems within a UGV. These may include batteries, fuel cells, or other power sources depending on the specific mission requirements.
4. Communication systems: Communication systems allow the UGV to transmit data and receive commands from remote operators or other vehicles in a networked environment. These may include wireless radios, satellite links, or other communication technologies.

5. Actuators: Actuators are responsible for controlling the movement of the UGV's various components such as wheels or manipulator arms. These may include electric motors, hydraulic actuators, or other types of actuators depending on the specific application.

   Each of these subsystems plays a critical role in enabling the UGV to perform its intended mission safely and effectively. By carefully selecting and integrating these subsystems based on specific mission requirements, you can create an efficient and effective UGV that meets your needs.

4. What are different missions a UGV is deployed for?

Answer: UGVs are deployed for a wide range of missions, each with its own specific requirements and challenges. Some of the most common missions for UGVs include:

1. Reconnaissance and surveillance: UGVs can be used to gather intelligence and provide situational awareness in a variety of environments, including urban areas, forests, and deserts. They can be equipped with sensors such as cameras, microphones, and chemical detectors to detect and identify potential threats.
2. Explosive Ordnance Disposal (EOD): UGVs can be used to safely locate, identify, and dispose of explosive devices such as landmines or Improvised Explosive Devices (IEDs). They can be equipped with manipulator arms or other tools to handle hazardous materials.
3. Search and rescue: UGVs can be used to search for survivors in disaster zones such as those hit by earthquakes or hurricanes. They can be equipped with sensors such as thermal cameras or microphones to detect signs of life.
4. Logistics support: UGVs can be used to transport supplies or equipment in environments where it is difficult or dangerous for humans to operate. They can be equipped with cargo bays or trailers to carry payloads.
5. Agriculture: UGVs can be used in agriculture for tasks such as planting seeds, spraying pesticides, or harvesting crops. They can be equipped with specialized tools such as seed drills or crop sensors.
6. Mining: UGVs can be used in mining operations for tasks such as drilling, blasting, or transporting ore. They can be equipped with specialized tools such as rock drills or conveyor belts.
7. Border patrol: UGVs can be used by border patrol agents to monitor remote areas along national borders. They can be equipped with sensors such as cameras or radar systems to detect potential threats.

   These are just a few examples of the many missions that UGVs are deployed for today. By understanding the specific requirements of each mission and selecting the appropriate UGV platform and subsystems, you can create an efficient and effective UGV that meets your needs.

5. Give different specifications of sensors used for different missions for a UGV.

Answer:

- 3DLidar (Velodyne VLP-16):
- Three sensors (one on either end of the front edge of the roof and one centrally located on the back edge of the roof)
- Tilted downwards by 10 degrees
- Used for environmental mapping and obstacle avoidance
- Colour camera (PointGreyBlackFly BFLY-PGE-23S6M-C):
- Three cameras (C1, C2, C3) mounted in a triangular configuration in the vertical plane facing forward
- One camera (C1) placed centrally at the front upper edge of the windscreen
- The other two cameras (C2 and C3) placed approximately at either end of the bottom edge of the windscreen with an adjustable baseline
- One camera each on either side of the platform (C4 and C5), towards the rear facing forward and down
- One camera mounted centrally facing the rear (C6)
- 8 mm and 12.5 mm lenses for cameras
- Global shuttering
- C4, C5, and C6 used primarily for teleoperation
- Minimum resolution of 640 x 480 pixels, with a minimum frame rate of 24 fps
- Thermal camera (XenicsServal 640 GigE):
- One camera each mounted centrally facing forward (T1) and rear (T2)
- Minimum resolution of 640 x 480 pixels and minimum frame rate of 24 fps
- 5.8 mm lens
- Used for teleoperation, especially in night-time conditions
- Time-of-flight camera (Fotonic E70):
- One camera each facing forward and rear
- Radars (UMRR-0C with type 128 antenna at 77 GHz):
- Multiple sensors mounted to provide a 120-degree horizontal field-of-view
- Range of at least 100 m ahead of the platform and 50 m behind the platform
- Capable of resolving in range, velocity and angular separation between objects
- Ultrasound sensors (MaxbotixMaxsonar E20XL):

- Multiple sensors mounted in a ring around the platform to ensure no dead zone beyond 1 m from the platform
- Max range of at least 7 m
- Coupled directly to the DBW subsystem for low-latency, reactive obstacle avoidance
- Integrated GNSS (INS, GNSS antennas, wheel speed sensor, RTK base station, and communication infrastructure):
- Dual GNSS receiver and IMU mounted inside the platform
- IMU centre of rotation centrally located above the rear axle
- Two GNSS antennas mounted on the roof as per recommended spacing
- Antenna to support at least GPS/GLONASS constellations (additionally IRNSS if available at the time of procurement)
- One wheel speed sensor mounted on one of the non-driven wheels
- Base station and communication infrastructure components for RTK to be set up and integrated with both receivers
- Online RTK solution to be established and integrated with the spatial system.

6. Give the technical specification of an embedded computer used to control an UGV.

Answer: The embedded computer used to control a UGV is a critical component of the vehicle's subsystems. It is responsible for controlling the various functions of the UGV, including navigation, communication, and sensor data processing. The technical specifications of an embedded computer used in a UGV will depend on the specific mission requirements and the capabilities of the other subsystems. Some key technical specifications to consider when selecting an embedded computer for a UGV include:

1. Processing power: The processing power of the embedded computer will determine how quickly it can perform complex tasks such as autonomous navigation or sensor data processing. This is typically measured in terms of clock speed (in GHz) or number of cores.
2. Memory capacity: The memory capacity of the embedded computer will determine how much data it can store and process at once. This is typically measured in terms of RAM (in GB).
3. Operating system: The operating system used by the embedded computer will determine what software can be run on it and how it interacts with other subsystems. Common operating systems used in UGVs include Linux and Windows.
4. Input/output interfaces: The input/output interfaces available on the embedded computer will determine how it communicates with other subsystems such as sensors or actuators. Common interfaces include USB, Ethernet, and serial ports.

5. Power consumption: The power consumption of the embedded computer will determine how much energy it requires to operate, which is an important consideration for battery-powered UGVs.
6. Environmental ruggedness: The environmental ruggedness of the embedded computer will determine how well it can withstand harsh conditions such as extreme temperatures or vibrations.
   When selecting an embedded computer for a UGV, it is important to carefully consider these technical specifications based on specific mission requirements and environmental conditions. By selecting an appropriate embedded computer that meets these requirements, you can ensure that your UGV is able to perform its intended mission safely and effectively

7. Give the functional requirement of the embedded software used in an UGV.

Answer: The embedded software used in a UGV is responsible for controlling the various subsystems of the vehicle, including navigation, communication, and sensor data processing. The functional requirements of this software will depend on the specific mission requirements and the capabilities of the other subsystems. Some key functional requirements to consider when developing embedded software for a UGV include:

1. Autonomous navigation: The embedded software should be able to control the UGV's movement autonomously, using inputs from sensors such as LiDAR or GPS to navigate through complex environments.
2. Sensor data processing: The embedded software should be able to process data from various sensors in real time, using algorithms such as object detection or image recognition to identify potential threats or obstacles.
3. Communication: The embedded software should be able to communicate with other vehicles or remote operators in a networked environment, using protocols such as TCP/IP or CAN bus.
4. Fault tolerance: The embedded software should be designed with fault tolerance in mind, so that it can continue operating even if one or more subsystems fail.
5. Security: The embedded software should be designed with security in mind, so that it can protect against cyberattacks or unauthorized access.
6. User interface: The embedded software should provide a user interface that allows operators to monitor and control the UGV's functions, using displays or input devices such as joysticks or touchscreens.
7. Power management: The embedded software should be able to manage power consumption efficiently, so that battery-powered UGVs can operate for extended periods of time without recharging.
   When developing embedded software for a UGV, it is important to carefully consider these functional requirements based on specific mission

requirements and environmental conditions. By designing an appropriate software architecture that meets these requirements, you can ensure that your UGV is able to perform its intended mission safely and effectively.

8. What are different user interface (UI0) components of a UGV?

Answer: The user interface of the Unmanned Ground Vehicle (UGV) will consist of two distinct components:

Platform-Mounted Display (PMD): This component will feature a sunlight-readable, touch-enabled display (at least 10”), and will be connected to the platform embedded computers through the KVM switch. The display positioning and mounting should be ergonomic.

Master Control Station (MCS): This component is required to be mobile and portable from one location to another rapidly. It will feature at least one sunlight-readable, touch-enabled display (at least 23), together with a keyboard, a mouse, and an embedded computer. It will also feature a console emulating the platform manual driving interface, with at least a steering wheel, accelerator, and brake pedals, together with seating for one operator. The positioning of the various components should be ergonomic. The operation of the steering wheel and pedals should closely correspond to that on the actual platform. The MCS should have an onboard rechargeable battery-based power pack for at least four hours of endurance with all components of the MCS operational. The charger for the power pack should be on board and should permit online charging through a standard 220 volt wall socket, while the loads are operational. The charger should be capable of optimizing the charging cycle of the attached battery as per the battery type. One set of spare batteries with charger and battery monitoring system needs to be kept operational, ready for any UGV operations. The packaging for the batteries must meet extant regulations governing the packaging for transport of such equipment, especially batteries. The MCS will communicate with the platform through a wireless communication link. Both the PMD and the MCS will feature a common Graphical User Interface (GUI).

REFERENCES

Apestigue V., Gonzalo A., Jiménez J.J., Boland J., Lemmon M., de Mingo J.R., García-Menendez E., Rivas J., Azcue J., Bastide L., Andrés-Santiuste N., Martínez-Oter J., González-Guerrero M., Martin-Ortega A., Toledo D., Alvarez-Rios F.J., Serrano F., Martín-Vodopivec B., Manzano J., LópezHeredero R., Carrasco I., Aparicio S., Carretero Á., MacDonald D.R., Moore L.B., Alcacera M.Á., Fernández-Viguri J.A., Martín I., Yela M., Álvarez M., Manzano P., Martín J.A., Del Hoyo J.C., Reina M., Urqui R., Rodriguez-Manfredi J.A., de la Torre Juárez M., Hernandez C., Cordoba E., Leiter R., Thompson A., Madsen S., Smith M.D., Viúdez-Moreiras D., Saiz-Lopez A., Sánchez-Lavega A., Gomez-Martín L., Martínez G.M., Gómez-Elvira F.J.,

Arruego I. Radiation and Dust Sensor for Mars Environmental Dynamic Analyzer Onboard M2020 Rover. Sensors (Basel). 2022 Apr 10;22(8):2907. doi: 10.3390/s22082907.

Boukoberine M.N., Zhou Z., Benbouzid M. A Critical Review on Unmanned Aerial Vehicles Power Supply and Energy Management: Solutions, Strategies, and Prospects. Applied Energy. 2019; 255:113823. ISSN 0306-2619. https://doi.org/10.1016/j.apenergy.2019.113823.

Chen, Xia et al. "PanoNet3D: Combining Semantic and Geometric Understanding for LiDAR Point Cloud Detection." *2020 International Conference on 3D Vision (3DV)* 2020: 753–761.

Fabris E.J., Sangalli V.A., Soares L.P., Pinho M.S. Immersive Telepresence on the Operation of Unmanned Vehicles. International Journal of Advanced Robotic Systems. 2021;18(1). doi: 10.1177/1729881420978544.

Ge S.S. Autonomous Mobile Robots: Sensing, Control, Decision Making and Applications. CRC Press; 2006 May 4, p. 584. ISBN 9781420019445. Retrieved 3 September 2016.

Guidi G., Borgarelli N., Malleret F. Benefits and Challenges of Electro-Mechanical Actuation on Aircraft Landing and Braking Systems – ATA32 EMAs. 2021. More Electric Aircraft Conference – Bordeaux, France.

Kandath H., Bera T., Bardhan R., Suresh S. Autonomous Navigation and Sensorless Obstacle Avoidance for UGV with Environment Information from UAV. *Second IEEE International Conference on Robotic Computing (IRC)*, Laguna Hills, CA, USA, 2018, pp. 266-269, doi: 10.1109/IRC.2018.00056.

Kozlov A., Kapralov F. Angular Misalignment Calibration for Dual-Antenna GNSS/IMU Navigation Sensor. Sensors. 2023; 23(1):77. https://doi.org/10.3390/s23010077

Liang Y., Zhao C., Yuan H., Chen Y., Zhang W., Huang J., Yu D., Liu Y., Titirici M., Chueh Y.L., Yu H., Zhang Q. A Review of Rechargeable Batteries for Portable Electronic Devices. InfoMat; 2019.

Majid M., Habib S., Javed A.R., Rizwan M., Srivastava G., Gadekallu T.R., Lin J.C-W. Applications of Wireless Sensor Networks and Internet of Things Frameworks in the Industry Revolution 4.0: A Systematic Literature Review. Sensors. 2022; 22(6):2087. https://doi.org/10.3390/s22062087

Nowakowski M., Kurylo J. Usability of Perception Sensors to Determine the Obstacles of Unmanned Ground Vehicles Operating in Off-Road Environments. Applied Sciences. 2023; 13(8):4892.

Parasuraman R., Caccamo S., Baberg F., Ogren P., Neerincx M. A New UGV Teleoperation Interface for Improved Awareness of Network Connectivity and Physical Surroundings. Journal of Human-Robot Interaction. 2017; 6:48–70. 10.5898/JHRI.6.3.

Rubio F., Valero F., Llopis-Albert C. A Review of Mobile Robots: Concepts, Methods, Theoretical Framework, and Applications. International Journal of Advanced Robotic Systems. 2019;16(2). doi: 10.1177/1729881419839596.

Tilbury D., Ulsoy A. Reliable Operations of Unmanned Ground Vehicles: Research at the Ground Robotics Reliability Center. 2010.

Chapter 12

Algorithms, control, and software used for UAS missions

OVERVIEW

This chapter discusses the algorithms, control, and application software used for accomplishment of the mission performed by unmanned autonomous systems. The two basic software cycles operated and controlled by software to smoothly conduct the mission of unmanned systems are discussed. Also discussed are some of the algorithms which are used for operating a swarm of unmanned systems such as swarms of Unmanned Aerial Vehicles, Unmanned Ground Vehicles, or Autonomous Underwater Vehicles.

12.1 INTRODUCTION

An Unmanned Autonomous System (UAS) can be modelled as a system of systems where there is a large percentage software interconnecting and controlling these subsystems in achieving the mission objectives. In other words, software forms a formidable and crucial part of the UAS which is not visible, unlike the sensors and the vehicle but performs and coordinates the entire mission of the UAS. The software employed to smoothly operate any UAS can be categorized under three broad categories, namely

(a) The Operating System (OS) or management software to control and manage the entire software stack and the sensors.
(b) The control software to control the sensors, payloads, and navigation sensors to navigate and collect data from the environment.
(c) The command and communication software, which communicates the command and control between various subsystems, the unmanned vehicle, and the mother station or ground station or recovery station.

Generally, a UAS (Unmanned Ground Vehicle, UGV; Unmanned Aerial Vehicle, UAV; or Autonomous Underwater Vehicle, AUV) is a Size, Weight, Area, and Power (SWAP)-constrained robotic platform which is constantly sensing the

DOI: 10.1201/9781003432685-12

environment and taking action as per the specific needs of the mission. Therefore, it is a real-time system with a platform fitted with sensors, actuators, and communication transmitters and receivers. The software which drives the UAS must have real-time response, a low memory footprint, and a low power-consuming system with high endurance and robustness to Electromagnetic Interference (EMI), Electromagnetic Compatibility (EMC), and environmental changes.

12.2 CHARACTERISTICS OF THE SOFTWARE USED IN THE UAS

The architecture of the software stack managing any unmanned system can be thought of as a loosely integrated components of three main components, the operating system, the middleware, and the mission control and application software. Further, these components are composed depending upon the mission objectives, type of autonomous craft or the vehicle, degree of autonomy of the UAS, and the persistence and endurance of the UAS. A generic block diagram is depicted in Figure 12.1 (Sanchez-Lopez et al., 2016).

UASs are becoming increasingly important in various industries, including the military, agriculture, and transportation. To ensure the smooth and safe operation of UASs, it is crucial to have reliable and efficient software that can manage the complex hardware and sensors involved in unmanned systems. In this regard, the following are some characteristics of the software used in UASs:

1. Robustness: UAS software must be designed to handle unexpected events and errors without compromising safety. The software should be able to recover from failures and continue operation even when a component fails.
2. Scalability: It must be scalable to handle different sizes of unmanned systems and workloads. The software should be able to accommodate changes in the number of sensors and payload capacity without impacting performance.
3. Modularity: It should be modular, with different components designed to work independently. This allows for easy integration of new sensors and payloads and makes it easier to replace faulty components.
4. Security: The software must be secure to protect against unauthorized access and prevent cyberattacks. The software should have authentication

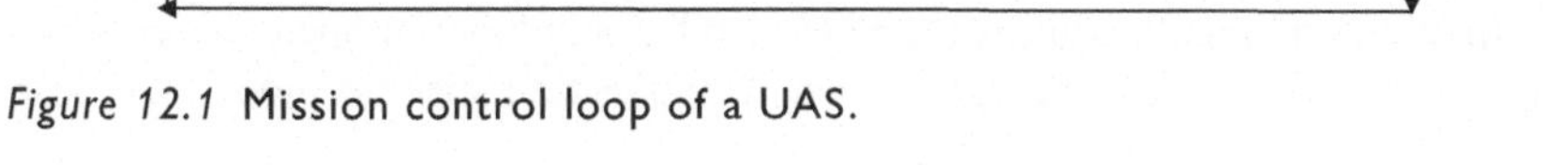

Figure 12.1 Mission control loop of a UAS.

and encryption mechanisms to ensure that only authorized personnel can access the system.

5. Real-time processing: There should be a capability of processing data in real time to support autonomous decision-making. This requires high-performance computing capabilities to analyse and process data quickly.
6. Interoperability: A UAS must be designed to work with other systems and platforms. The software should be able to communicate with different types of ground control stations and other unmanned systems to support coordinated operations.

12.3 OPERATING SYSTEMS USED IN UNMANNED SYSTEMS

The OS used in unmanned systems manages all other software and provide a real-time platform to execute and manage other software modules in a conflict-free manner. The persistence (duration of program in execution) of the OS is the entire duration of the mission. Therefore, the essential requirements and characteristics of an OS-managing UAS are (a) Low memory foot print, (b) Low power consumption, (c) High persistence, and (d) Real-time response. Often, the combination of these specifications are called the SWAP constraint. There are many SWAP OS with low-memory footprint, high persistence, and interrupt handling capability such as Android, WinCE, and the Robotic Operating System (ROS) used especially for robotic systems, and small form-factor and application-specific devices such as mobile computing platforms and cellular phones (Mohsan et al., 2023).

12.3.1 Operating system and middleware used by UASs

UASs are highly complex systems that require advanced software and middleware to operate effectively. One of the most widely used middleware in UASs is the ROS. The ROS provides a flexible framework for building complex robotic systems and has been used extensively to implement autonomous navigation on tracked and wheeled UGVs, and other smaller mobile robotic platforms.

The ROS provides a wide range of core features, including topics, services, actions, parameters, dynamic parameter reconfiguration, logging, diagnostics, and node lets, that can be used to build mission-specific software packages. The catkin and colcon build systems are employed to build these software packages, which can be written in C++, Python, and Java. Additionally, the ROS also supports integration with web interfaces through the rosbridge and rosjs client library packages.

In addition to its flexibility, the ROS also has a number of open-source packages available that can be used to build application-specific modules. These packages include sensor drivers for mono/stereo colour/thermal cameras, GNSS/INS, 2D/3D Light Detection and Ranging (LiDAR), and depth cameras. The navigation and MoveIt stacks available in the ROS are also very useful for implementing navigation and mobility-related functions. Other important packages such as robot_localization, sbpl_planner, teb_local_planner, and slam_toolbox can be used for Simultaneous Localization and Mapping (SLAM), path planning, and localization of the platform (Ashraf et. al., 2020).

To support development and debugging, the ROS provides both Client Library Interface (CLI) and Graphical User Interface (GUI) tools, including rostopic, rosservice, rviz, rqt_graph, rqt_console, and rqt_reconfigure. These tools are used extensively for introspection and debugging during UAS development.

12.3.2 Mission control software of UASs

Software plays a crucial role in the success of missions involving UAS. The mission control software which is persistent software is a vital component of any unmanned system, imparting liveliness, autonomous capability and intelligence to the mission activities. Since the activities of a mission are real time, most of the decisions regarding movement of the unmanned system and the sensor operations are coordinated by the Mission Control Software (MCS). The unmanned system performing the mission activities needs to sense the environment, which is an unknown factor, and take autonomous decisions regarding its environment-sensing movement as per built-in navigation algorithm and mission requirements.

Once the mission is on and the UAS is in action, there are two activities which are performed continuously, namely, the sensing of the environment through various sensors and computing the real-time scenario for topological representation of the environment in the form of location of obstacles, objects, slope, aspect, curvature, and other geometric properties which are required in order to take decisions regarding movement of the UAS. The other persistent task done by the UAS is navigation, that is, navigating to the destination or the point of objective as decided by the mission. Navigation is a continuous process of comparing the existing state of the UAS with respect to the current environment. It decides what are obstacles to the navigation and what are aids to the navigation, compares these with the preexisting path, computes deviation, if any, and finally takes a decision so as to avoid obstructions and make an incremental approach to the destination or objective. Therefore, these two mission tasks persist and manifest as a loop in the entire life of the mission. These two software loops are depicted in Figure 12.2 (Melo & Lee, 2018).

The mission control loop of a UAS is depicted in Figure 12.1. It involves four basic steps, namely, localization, perception, path planning and motion control.

A. Localization: The location of a UAS is critical for its safe and effective operation. The first task that a UAS performs upon deployment is to determine its location. This is done by obtaining the (lat., long.) coordinates for a UGV, the (lat., long., altitude) coordinates for a UAV, and the (lat., long., and depth) coordinates for an AUV. The most common method for obtaining the UAS location is by using Global Navigation Satellite Systems (GNSS) sensors, such as GPS/GLONASS/IRNSS. These sensors produce output in an NMEA format, which the programme reads and decodes to obtain the (lat., long., altitude) coordinates, as well as the speed and time of the UAS.

However, there are situations where the GNSS system is unavailable or in a GNSS-denied environment. In such cases, dead reckoning or the Inertial Navigation System (INS) can be used to compute the location of the UAS. Dead reckoning involves estimating the current location of the UAS by using its last known position, along with its speed and direction of motion. The INS, in contrast, uses readings from the speedometer, accelerometer, and angular motion sensors to interpolate the current location of the UAS. These interpolated values are then augmented with the computed location and interpolated location to obtain the location of the UAS.

It is important to note that determining the location of a UAS can be challenging in certain situations. For example, GNSS signals may be blocked or degraded by natural or man-made structures such as thick canopies, tall buildings, or underwater environments. In such cases, UAS developers need to consider alternative methods for determining the location of their systems, such as dead reckoning or INS. By understanding these challenges and alternative methods, UAS developers can design systems that are capable of operating effectively in a variety of environments and situations.

B. Perception: The second task a UAS performs is perceiving the surrounding environment through the attached sensors and generating a cost map of the environment. The perception by UAS is performed through sensors such as the proximity sensor, LiDAR, sonar, and radar, or through vision-based sensors such as mono-focal or bi-focal optical sensors. The UAS constantly senses the environment through these sensors and computes the distance from the current location to various obstacles. It maps the slope, aspect, and curvature of the terrain under the field of view of these sensors and constructs a map of the environment, which is known as a cost map of the environment. A cost map usually classifies the terrain into three categories of going conditions for the UAS viz. good goable, partial goable, or no-goable. The classified terrain map is known as going map having classified the map into three categories good goable, partial goable, or no-goable.

Performing this localization of the UAS and construction of the environmental cost map simultaneously is known as SLAM. This is a perpetual or continuous loop throughout the mission of a UAS.

C. Path Planning: Planning the path for a UAS is a critical task that must be performed accurately to ensure a successful mission. The path planning is determined by considering factors such as the type of UAS, the mission objectives, the environmental conditions, and the obstacles in the UAS's path. The path planning algorithm must also take into account the energy consumption and the time required to complete the mission.

In recent years, advanced path planning algorithms have been developed that employ machine learning and artificial intelligence techniques. These algorithms enable the UAS to make real-time decisions based on changing conditions in the environment. Machine learning algorithms can be used to classify obstacles, terrain types, and other environmental factors, and employ this information to plan the most efficient and safe path for the UAS.

Furthermore, in addition to the conventional path planning techniques like A-star, Dijkstra, and Ant colony optimization, more advanced techniques like Rapidly-exploring Random Trees (RRT), Probabilistic Roadmaps (PRM), and potential field-based methods are also used to plan the path of UAS.

Lastly, path planning must take into consideration the legal and ethical aspects of the mission. In some cases, UAS are subject to airspace regulations that dictate their flight paths and altitudes. It is important for UAS operators to adhere to these regulations to avoid any legal repercussions. Additionally, path planning must take into account the privacy and security concerns of individuals in the area of operation to prevent any violations of their rights.

D. Motion Control: The fourth step for a UAS is to control and manage the speed and motion of the wheel actuators controlling its steering and wheels. This is known as motion control of the UAS.

The above four steps are often known as the mission control loop of a UAS and are depicted in Figure 12.1.

To prepare a cost map of the environment and update the next location of the UAS, the above four activities are performed continuously and often concurrently. This imparts autonomous capability and real-time control to the vehicle, its sensors, communications, and kinematic. Therefore, the software that performs these activities should have real-time control of the vehicle, its sensors, communications, and kinematic. The software that performs these activities is real time in nature, often implemented as firmware embedded into the hardware platform of the unmanned system for real-time response. Besides this, the software needs to be robust, fail-safe, and trustworthy with known bounds of accuracy. Therefore, necessary fail-safe methods are embedded into the MCS to accomplish the UAS mission (Wiggers, 2020).

The software governing the operations of the Unmanned Autonomous System (UAS) has multiple real-time loops persistently running during the mission. These are

a. The mission management-related software loop.
b. The loop managing the operations of payload sensors.
c. The loop managing the kinematic of the AUV/UAV/UGV.
d. Telemetry or the communication between the GCS and the AUV/UAV/UGV for command, control, and communication within a man-in-loop.

12.4 COMMUNICATION BETWEEN UAS AND GCS

The communication between the GCS and the unmanned vehicle is not a conventional communication channel. The communication link between the GCS and the UAV is characterized by low bandwidth, as dyadic, fragile (frequent link failure), heterogeneous, and often restful in nature. Therefore, it is not a real-time communication. The scenario of communication between the mother station and a UAS is through an anisotropic medium or heterogeneous medium such as air-water (deep sea), air-to-air, ground-to-satellite, and so forth. The communication between UAS and GCS is mostly non-Line-of-Sight (non-LoS), ground to air with intermittent link failure. Therefore, the communication link and its liveliness are depicted using a digital map or electronic navigation chart or air navigation chart. This plots the LoS between the craft and the GCS, marking it green if it is available, and red if not available. The communication between the GCS and the UGV is often non-LoS, wireless, and frequently jittery.

Following are some of the common communication impairments in missions that are often found between the GCS and the autonomous vehicle:

1) Non-LoS
2) Long-haul wireless
3) Intermittent link failure
4) Band limited and Error prone

Therefore, the communication protocol should be robust, opportunistic, REST, SWAP, and heterogeneous (medium of the link can be partly air, water, wired, and space).

A communication protocol where the information is Stored and Forward (SnF) is important for such communication. The communication protocol to receive, store, and forward mission-critical non-real-time data and messages over intermittent links has to be SnF and Rest in design. The approach enables reliable data (mission critical) communication between intermittently connected end hosts (UAS and GCS) across the network.

To design and craft this kind of communication channel is a challenge, with heterogeneous communication links between the unmanned systems with the mother station.

12.4.1 Sensor control and alignment algorithms

More commonly known as a **driver**, a **device driver,** or a **hardware driver** is a group of files that enable one or more hardware devices to communicate with the computer's operating system. Without drivers, the computer could not send and receive data correctly to hardware devices, such as a printer.

If the appropriate driver is not installed, the device may not function properly, if at all. With some devices, the device may work, but all of its features may not work. For example, a computer mouse usually works without drivers, but if it has more buttons than the traditional mouse, those extra buttons will not work until the drivers are installed.

What are the sensor controller source files used for?
These source files contain the sensor controller firmware image and associated definitions, and generic functions that allow the system CPU application to control the sensor controller and exchange data. The sensor controller is a small CPU core that is highly optimized for low power consumption and efficient peripheral operation

What is the sensor controller?
The sensor controller is a small CPU core that is highly optimized for low power consumption and efficient peripheral operation. The sensor controller is located in the CC26xx/CC13xx auxiliary (AUX) power/clock domain, and can perform simple background tasks autonomously and independently of the system CPU and the MCU domain power state.

Free **Sensor Control** software powers all Spectral Engines' products by giving a true plug-and-play experience with your spectral measurements. All the parameters of the spectral sensors and light sources can be adjusted. Freely programmable spectral wavelength points of the spectral sensors and basic analysis functions, along with versatile data-saving options give you powerful tools to study your samples.

SDK Software Development Kit
Our Software Development Kit (**SDK**), makes it easy to start your own design and software programming. This SDK is specially developed for system integrators who like to get access to microcontrollers and design their own software and user interface. SDK provides the communication interface description which allows you to use your preferred programming language. Example code included in the SDK gives you good insight into how to control the sensor. With these tools, you can create your own interactive or even fully automated measurement system.

12.4.2 Device drivers and sensor-controlling software

Device drivers are software components that facilitate communication between the hardware devices (such as sensors, actuators, and other peripherals) and the operating system. In the case of UASs, device drivers are responsible for controlling and managing the sensors and other hardware components on board the aircraft. The sensor-controlling software is another important component that plays a critical role in the communication between the UAS and the GCS. This software is responsible for processing the data from the sensors and converting it into a format that can be easily understood by the GCS. It also helps in filtering and smoothing the data to remove any noise or inaccuracies that may arise due to the environment or the hardware (Kangunde et al., 2021).

In UASs, there are different types of sensors that are used to gather information about the environment and the aircraft's state. These sensors include GPS, altimeters, accelerometers, gyroscopes, magnetometers, cameras, LiDAR, and many others. Each sensor has its own unique characteristics, data format, and communication protocol. Therefore, the device drivers and sensor-controlling software must be designed to handle these differences and ensure compatibility between the sensors and the GCS. Moreover, the communication between the UAS and the GCS is critical for the operation of the UAS, and any loss or interruption in communication can have serious consequences. The device drivers and sensor-controlling software must be designed to handle communication errors and ensure that the data is transmitted reliably and efficiently.

12.4.3 Navigation algorithms used in UASs

One of the most commonly used navigation algorithms for UASs is the Global Positioning System (GPS). GPS uses a network of satellites to provide location and time information to the UAS. This information is then used by the navigation system to calculate the UAS's position, velocity, and orientation. GPS is highly accurate and reliable, making it an ideal navigation solution for the UAS. Another important navigation algorithm for the UAS is the INS. The INS uses accelerometers and gyroscopes to measure changes in the UAS's velocity and orientation. By integrating these measurements over time, the INS can determine the UAS's position relative to its starting point. The INS is particularly useful in environments where GPS signals may be weak or disrupted, such as urban areas or inside buildings.

Communication between the UAS and the GCS is also a critical component of navigation. The UAS must be able to transmit real-time information about its location, speed, and status to the GCS, while the GCS must be able to send commands to the UAS to adjust its flight path or mission objectives. To facilitate this communication, many UASs use Radio Frequency (RF) transceivers to send and receive data between the UAS and the GCS. One important consideration for communication between the UAS and the GCS is the bandwidth and reliability of the RF link. UASs typically transmit large amounts of data, including video streams, telemetry data, and sensor data. To ensure that this data is transmitted reliably, UASs often use error-correcting codes and data compression techniques to reduce the amount of data that needs to be transmitted. Additionally, UASs may use multiple RF links or alternate between different frequencies to improve the reliability of the link.

12.4.4 Navigation of UASs

Algorithms for collision-free navigation (Paloma et al., 2020) of mobile robots in complex cluttered environments is a challenging task. There are a range of techniques related to navigation of unmanned vehicles through unknown environments with obstacles, especially those that rigorously ensure collision avoidance (given certain assumptions about the system). This topic continues to be an active area of research, and we highlight some directions in which available approaches may be improved. The chapter discusses models of the sensors and vehicle kinematics, assumptions about the environment, and performance criteria. Methods applicable to stationary obstacles, moving obstacles, and multiple vehicle scenarios are all reviewed. In preference to global approaches based on full knowledge of the environment, particular attention is given to reactive methods based on local sensory data, with a special focus on recently proposed navigation laws based on model predictive and sliding mode control.

Navigation of unmanned vehicles is a classic research area in robotics and unmanned systems. There is a lot of research literature with different approaches available in the open literature and proprietary research as well. Both single and multiple coordinated unmanned vehicles offer great perspectives in many applications, such as industrial, official, and agricultural automation; search and rescue; and surveillance and inspection etc. In all these cases, navigation involves a series of common problems, with collision avoidance in some form being almost universally needed. This section we mention methods relating to navigation of unmanned vehicles as applicable to collision avoidance. Among various methods we emphasise those which are capable of rigorous collision avoidance, that is, for which it is rigorously proven that collisions never occur, provided that some technical assumptions are fulfilled. Provable collision avoidance is a highly desirable trait in navigation systems since it shows that the system safely operates under a broad range of conditions, rather than just those considered during testing. Overall, navigation systems underlaid by more general assumptions about the problem would be considered superior. Some examples of assumptions are listed as follows:

- Vehicle models vary in complexity from velocity-controlled linear models to realistic car-like models. For example, collision avoidance for velocity-controlled models is simpler since the vehicles can halt instantly if required; however, this is physically unrealistic. This means a more complex model that better characterizes the vehicle is desirable for use during analysis.

  Different levels of knowledge about the obstacles and other vehicles are required by different navigation strategies. This ranges from abstracted obstacle set information to the actual nature of realistic noisy sensor data obtained from range-finding sensors. In addition to the realism of the sensor model, the actual sensing requirements vary significantly between approaches in some cases, methods that only utilize limited information (such as the minimum distance to the obstacle) are required.
- Different assumptions about the shape of static obstacles have been proposed. To ensure correct behaviour when operating near an obstacle with limited information available, it is often necessary to presume smoothness of the obstacle boundary. This means that approaches with more flexible assumptions about obstacles would be more widely applicable to real-world scenarios.
- Uncertainty is always present in real robotic systems, and proving collision avoidance under an exact vehicle model will not always imply that the associated correct behaviour will be exhibited when implemented. To reflect this, assumptions can be made describing bounded disturbance from the nominal model, bounded sensor errors, and the presence of communication errors. A review of the types of uncertainty present in vehicle systems is available.

12.5 WHAT IS SLAM AND HOW DOES SLAM WORK?

SLAM is a method used for autonomous vehicles that lets you build a map and localize your vehicle on that map at the same time. SLAM algorithms allow the vehicle to map out unknown environments. Engineers use map information to carry out tasks such as path planning and obstacle avoidance.

To build the map of the environment, the SLAM algorithm incrementally processes the LiDAR scans and builds a pose graph that links these scans. The robot recognizes a previously visited place through scan matching and may establish one or more loop closures along its moving path. The SLAM algorithm utilizes the loop closure information to update the map and adjust the estimated robot trajectory.

A map generated by a UAS or robot using SLAM algorithm is a computational problem of constructing or updating a map of an unknown environment while simultaneously keeping track of an agent's location within it (Newman & Ho, 2005).

Broadly speaking, there are two types of technology components used to achieve SLAM. The first type is sensor signal processing, including the front-end processing, which is largely dependent on the sensors used. The second type is pose-graph optimization, including the back-end processing, which is sensor-agnostic.

12.5.1 Visual SLAM

As the name suggests, visual SLAM (or vSLAM) uses images acquired from cameras and other image sensors. Visual SLAM can use simple cameras (wide angle, fish-eye, and spherical cameras), compound eye cameras (stereo and multi cameras), and RGB-D cameras (depth and ToF cameras).

Visual SLAM can be implemented at low cost with relatively inexpensive cameras. In addition, since cameras provide a large volume of information, they can be used to detect landmarks (previously measured positions). Landmark detection can also be combined with graph-based optimization, achieving flexibility in SLAM implementation.

Monocular SLAM is when vSLAM uses a single camera as the only sensor, which makes it challenging to define depth. This can be solved either by detecting AR markers, checkerboards, or other known objects in the image for localization, or by fusing the camera information with another sensor such as Inertial Measurement Units (IMUs), which can measure physical quantities such as velocity and orientation. Technology related to virtual SLAM includes Structure from Motion (SfM), visual odometry, and bundle adjustment.

Visual SLAM algorithms can be broadly classified into two categories. Sparse methods match feature points of images and use algorithms such as PTAM and ORB-SLAM. Dense methods use the overall brightness of images and employ algorithms such as DTAM, LSD-SLAM, DSO, and SVO.

12.5.2 LiDAR SLAM

LiDAR is a method that primarily uses a laser sensor (or distance sensor).

Compared to cameras, ToF, and other sensors, lasers are significantly more precise, and are used for applications with high-speed moving vehicles such as self-driving cars and drones. The output values from laser sensors are generally 2D (x, y) or 3D (x, y, z) point cloud data. The laser sensor point cloud provides high-precision distance measurements, and works very effectively for map construction with SLAM. Generally, movement is estimated sequentially by matching the point clouds. The calculated movement (travelled distance) is used for localizing the vehicle. For LiDAR point cloud matching, registration algorithms such as Iterative Closest Point (ICP), and Normal Distributions Transform

(NDT) algorithms are used. 2D or 3D point cloud maps can be represented as a grid map or voxel map.

On the other hand, point clouds are not as finely detailed as images in terms of density and do not always provide sufficient features for matching. For example, in places where there are few obstacles, it is difficult to align the point clouds, and this may result in losing track of the vehicle location. In addition, point cloud matching generally requires high processing power, so it is necessary to optimize the processes to improve speed. Due to these challenges, localization for autonomous vehicles may involve fusing other measurement results such as wheel odometry, GNSS, and IMU data. For applications such as warehouse robots, 2D LiDAR SLAM is commonly used, whereas SLAM using 3D LiDAR point clouds can be used for UAVs and automated driving (Mur-Artal et al., 2017).

12.5.3 GNSS-based navigation

GNSS plays a critical role in the navigation of the UAS. GNSS-based navigation allows the UAS to determine its precise location, velocity, and orientation in real time. GNSS uses a constellation of satellites orbiting the Earth to provide continuous and accurate position and timing information to GNSS receivers on the UAS.

The GNSS receiver on the UAS computes the location of the UAS by measuring the time delay between the transmission of signals from the satellites and their reception at the receiver. By measuring the time difference between signals from multiple satellites, the receiver can determine its position using trilateration. In addition to position, the GNSS receiver can also calculate the velocity and orientation of the UAS.

GNSS-based navigation is essential for UAS operations, especially for Beyond Visual Line of Sight (BVLoS) operations. However, GNSS signals can be easily disrupted by environmental factors such as multipath, signal blockage, or interference. As a result, GNSS navigation systems are augmented with other navigation sensors such as INS to provide more accurate navigation solutions.

12.5.4 Navigation in GNSS-denied situations

GNSS is a crucial tool for navigation, guidance, and control of UASs. However, in certain situations, GNSS signals may be denied or degraded due to factors such as signal jamming, multipath interference, or atmospheric conditions. In such GNSS-denied situations, UASs must rely on alternative methods of navigation to ensure safe and efficient flight operations (Sabatini et al., 2017).

One of the most commonly used methods for navigation in GNSS-denied situations is the use of an INS. An INS utilizes accelerometers and gyroscopes to determine the UAS's position, velocity, and attitude based on the integration of the acceleration and angular rate measurements. However, an INS has a significant disadvantage of accumulating errors over time, which may lead to degraded navigation accuracy.

An INS consists of accelerometers and gyroscopes that measure the UAS acceleration and angular velocity, respectively. By integrating these measurements over time, the UAS system can determine the UAS position and orientation relative to its starting point. INS systems can provide accurate navigation information for short periods, but errors can accumulate over time, leading to drift from the desired path.

To mitigate the effects of drift, UASs can use Visual Odometry (VO) and SLAM techniques. VO involves using cameras to track visual features in the environment and estimate the motion of the UAS. SLAM involves constructing a map of the environment using sensor data and estimating the position of the UAS relative to that map. Both VO and SLAM can improve navigation accuracy, but they require sufficient visual information, and the algorithms can be computationally intensive.

Another method for navigation in GNSS-denied situations is the use of a Visual Navigation System (VNS). A VNS relies on the UAS's on-board camera to capture images of the surrounding environment, and then utilizes image processing algorithms to extract features such as landmarks, buildings, or natural features. These features are then used to estimate the UAS's position, velocity, and attitude. A VNS has the advantage of being accurate, robust, and cost-effective, but may be limited in certain environments with poor lighting or low feature density.

Another approach to navigation in GNSS-denied situations is to use RF sensors such as radio altimeters, radar, and laser range finders. These sensors can measure the distance between the UAS and the ground or obstacles, providing information about the altitude and position of the UAS. However, RF sensors can be affected by interference and clutter, and their accuracy can be affected by the environment. Also RF navigation systems use signals transmitted by ground-based beacons or transmitters to determine the UAS's position. These signals can be either omnidirectional or directional, and the UAS's receiver can measure the time of arrival and phase difference of the signals to determine its position. RF navigation systems have the advantage of being robust to GNSS jamming and interference, but may have limited coverage and require significant infrastructure (Roumeliotis & Johnson, 2002).

Finally, UASs can use a combination of these techniques to achieve robust navigation in GNSS denied situations. For example, an INS can provide a starting point for VO and SLAM algorithms, which can then refine the position estimate. RF sensors can provide additional information about the environment and improve accuracy. The choice of navigation strategy depends on the specific application and the level of accuracy required.

i. **Path Planning Algorithm RRT Dijakstra Algorithm, A-Star Algorithm, Bellman Ford Algorithm (AI -> Reinforcement Learning)** (Stefano et al., 2001)

Path planning algorithms are essential for the navigation of UASs and other autonomous vehicles. These algorithms help in finding the optimal path from the current location of the UAS to the desired destination while avoiding obstacles and other potential hazards. Some popular navigation algorithms used in UASs are RRT, Dijkstra, A*, and Bellman Ford.

A. **Rapidly-exploring Random Tree (RRT) Algorithm**

The RRT algorithm is a sampling-based algorithm that explores the search space by randomly building a tree of connected nodes. It gradually grows this tree, taking random samples in the search space, until it reaches the goal node. The RRT algorithm is known for its scalability and efficiency, and it is commonly used in navigation problems.

B. **Dijkstra Algorithm**

The Dijkstra algorithm is a classical algorithm for finding the shortest path between two points in a graph. It works by maintaining a set of visited nodes and a set of unvisited nodes, and then iteratively choosing the unvisited node with the smallest distance from the starting node. The algorithm guarantees that it will find the shortest path, but it may take a long time to compute on larger graphs.

C. **A-Star Algorithm**

The A* algorithm is a heuristic search algorithm that is widely used in robotics and game development. It combines the advantages of the Dijkstra algorithm and heuristic functions to find the shortest path more efficiently. The A* algorithm uses a heuristic function to estimate the distance from the current node to the goal node. This helps to guide the search towards the goal, making it faster than Dijkstra's algorithm.

D. **Bellman-Ford Algorithm**

The Bellman-Ford algorithm is a dynamic programming algorithm used to find the shortest path between two nodes in a weighted graph. It can handle graphs with negative weight edges and is often used in routing protocols. However, it is slower than Dijkstra's algorithm and A* algorithm.

Recently, reinforcement learning-based algorithms have also gained popularity in path planning problems (Chen et al., 2014). These algorithms can learn optimal policies through trial and error without explicit knowledge of the environment. These algorithms are capable of learning from experience and adapt to changing conditions, making them useful in complex environments where traditional algorithms may not perform well

ii. **Computer Vision Algorithms (Localization, Perception, and Navigation Algorithms)** (Engel et al., 2014)
Computer vision algorithms play a vital role in the development of navigation algorithms for autonomous systems, including UAVs, self-driving cars, and mobile robots. These algorithms allow the system to perceive and understand its environment, localize itself within the environment, and navigate to a desired destination.

Localization algorithms use computer vision techniques to determine the system's position within a map or a previously explored environment. These algorithms use various sensors, including cameras, LiDAR, and radar, to extract visual features or 3D point clouds of the surrounding environment. The system can then match these features with a pre-built map to estimate its position accurately. Localization algorithms play a crucial role in autonomous navigation, allowing the system to operate safely and effectively in real-world environments (Samadzadegan, 2021).

Perception algorithms enable the system to interpret the environment by identifying and classifying objects, such as pedestrians, cars, traffic signs, and obstacles. These algorithms use machine learning techniques to train a model that can recognize and label different objects in real time. Perception algorithms are essential for autonomous vehicles, enabling them to detect and react to their surroundings, making critical decisions in real time.

Navigation algorithms use the output from localization and perception algorithms to plan a safe and efficient trajectory towards a desired destination. These algorithms take into account the system's current position, the destination, and any obstacles or constraints in the environment. Navigation algorithms also consider various factors such as speed, acceleration, and manoeuvrability to ensure the system can reach its destination while adhering to safety regulations.

12.6 OVERALL COMMAND AND CONTROL LOOP MANAGING THE UAS

The MCS works in multiple loops or loops-inside-loop or human-in-hardware-loop, or human-in-software-loop methods. The degree of independence of the MCS in the AUS mission determines its autonomous nature.

The man-in-loop operations are discussed here. The computer vision and AI algorithms that make the Unmanned System (US) an intelligent platform and brings in autonomous functioning of the craft as required separately for different AUSs.

The MCS is the live control mechanism which makes any unmanned system autonomous, intelligent, robust, and fail-safe. Therefore, the usefulness of the MCS decides the success of any unmanned system; in fact, it is akin to the pulse of the human body.

According to the degree of control of software, a UAS mission can be categorized as fully automatic, semi-automatic, or controlled by a human pilot on ground running the mission remotely. These categorizations of a UAS mission are depicted in Figure 12.2.

Base station/mother station/ground control station, with a ground pilot.

Mission Initialization:

I. Initialization of sensors
II. Communication check
III. Population check
IV. Control and avionic check
V. Initialization of duration of the mission

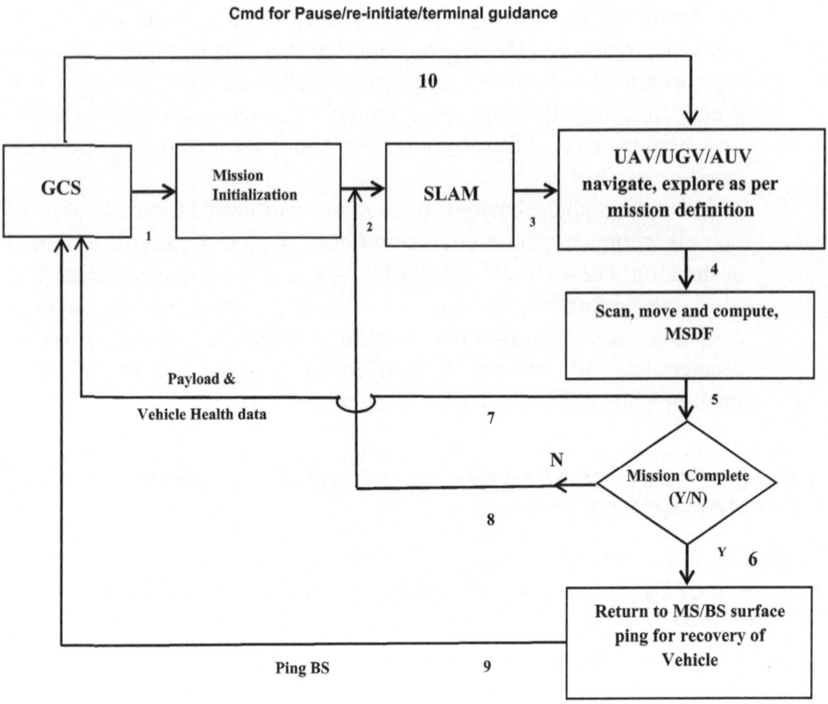

Figure 12.2 UAS mission control loop.

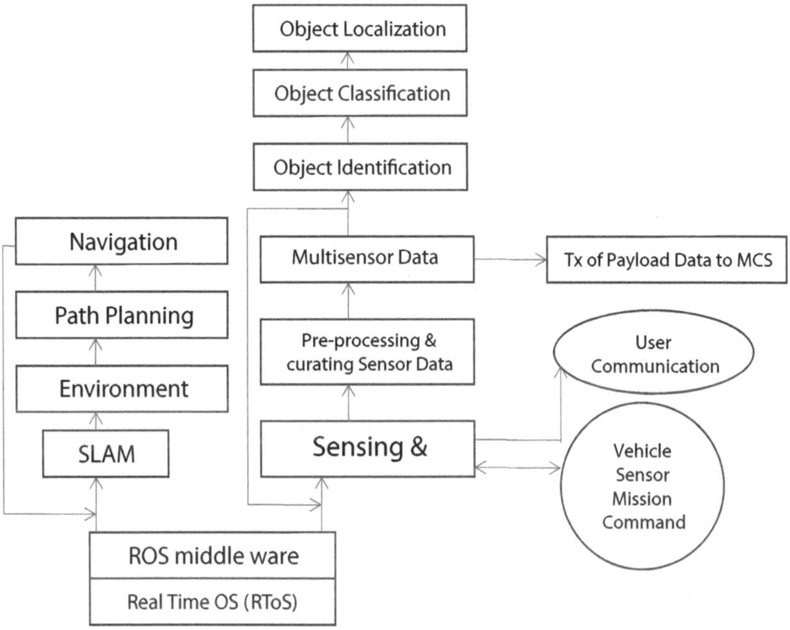

Figure 12.3 The processing cycles of a UAS mission.

The overall command and communication loop managing the UAS mission is depicted in the figure. If the mission follows the loop (1-2-3-4-5-6-7/8) during a mission, then it can be called a fully autonomous mission (Kwon, 2020). However, if the mission follows (1-2-3-4-5-7-10), then the mission can be categorized as a semi-autonomous mission (Figure 12.3).

QUESTIONS AND ANSWERS

1. What are different categories of software employed in operating a UAS, UAV, UGV, and AUV? How are they similar and what are their specific differences?

Answer: There are three broad categories of software used in operating UASs, including (a) the operating system for managing the software stack and sensors; (b) the control software for controlling the sensors, payloads, and navigation sensors; and (c) the command and communication software for communication between subsystems and ground stations. These software categories are similar in that they are geared towards achieving mission objectives in real time, but their differences lie in their specific functions, such as managing the software stack or controlling payloads.

2. What is a UAS from the perspective of a robotics system? What are its hardware and software components?

Answer: From the perspective of a robotics system, a Unmanned Autonomous System (UAS) is a system of systems, comprising of sensors, actuators, communication transmitters, and receivers. The software stack managing any unmanned system can be thought of as loosely integrated components of three main components: the operating system, the middleware, and the mission control and application software.

3. What are the generic characteristics of the software employed in a UAS?

Answer: The software employed in a UAS must be robust, scalable, modular, secure, and have real-time processing capability. It must be able to handle unexpected events and recover from failures, accommodate changes in sensor and payload capacity, work independently, have mechanisms to prevent cyberattacks and authenticate authorized personnel, and process data quickly to support autonomous decision-making.

4. What is a middleware software in UASs? Why is middleware required? What are the functions of middleware?

Answer: Middleware is software that acts as a bridge between different software components in a UAS. It provides a standard communication protocol for exchanging data between components, laying the foundation for integration. Middleware is required to facilitate communication between hardware and software components, ensuring that messages are delivered reliably and efficiently. Some functions of middleware include message routing, data conversion, and error handling.

5. What is a mission control software in a UAS? What are the functional specifications of a mission control software?

Answer: A mission control software (MCS) is the live control mechanism that makes an unmanned system autonomous, intelligent, robust, and fail-safe. The usefulness of the MCS decides the success of any unmanned system, and it works in multiple loops managing the UAS mission. The functional specifications of the MCS include robustness, scalability, modularity, security, and real-time processing.

6. What are the basic components of a mission control software in a UAS?

Answer: The basic components of a mission control software in a UAS are planning and scheduling, vehicle control, sensor management, data processing, and communication. These components work together to plan and execute mission objectives through autonomous or manual control.

7. What are the communication links between a UAS and a GCS? What are the common communication impairments between a GCS and a UAS?

Answer: Communication links between a UAS and a GCS are typically made through RF transceivers. Common communication impairments include non-LOS, long-haul wireless, intermittent link failure, and band-limited and error-prone connections. These impairments can be overcome through the use of a robust, opportunistic, REST, SWAP, and heterogeneous communication protocol, which incorporates Store and Forward (SnF) to enable reliable communication between intermittently connected end hosts.

8. What is SWAP? How is a UAS a SWAP device? How is the software employed in a SWAP device different from that of a software employed in a normal robotics device?

Answer: SWAP stands for Size, Weight, Area, and Power, which refers to the physical constraints and limitations of the robotic platform. A UAS is a SWaP device as it is a small, lightweight, and energy-efficient robotic system that can operate for extended periods of time while remaining within weight and power limits. The software employed in a UAS must be designed to have a low memory footprint, low power consumption, real-time response, and high endurance and robustness to environmental changes like EMI and EMC.

9. Give the software stack used in managing the mission of a UAS. Describe the functions of each of the software components used in UAS.

Answer: The software stack used in managing the mission of a UAS consists of the Operating System (OS), the middleware, and the mission control and application software. The OS controls and manages the entire software stack and the sensors; the middleware provides communication between different software components and interfaces with the hardware; and the mission control and application software controls the sensors, payloads, and navigation sensors to collect data and drive decision-making.

10. What is sensor control software in a UAS? How are the UAS sensors controlled and why is this necessary?

Answer: Sensor control software in a UAS is responsible for processing the data from on-board sensors and converting it into a format that can be easily understood by the GCS. It also helps in filtering and smoothing the data to remove any noise or inaccuracies that may arise due to the environment or hardware. This software is necessary to ensure compatibility between sensors and the GCS, and to handle communication errors to ensure data is transmitted reliably and efficiently.

11. What is SLAM? How is SLAM achieved in UAS systems? Why is SLAM necessary?

Answer: SLAM is a technique used in autonomous systems to map an unknown environment while simultaneously determining the system's location within it. In a UAS, SLAM is achieved by incrementally processing LiDAR scans and building a pose graph that links the scans. SLAM is necessary in a UAS to enable the system to operate autonomously while navigating and avoiding obstacles.

12. What are different types of SLAM? Where are these SLAM techniques used in UAS missions?

Answer: There are different types of SLAM techniques, including EKF-based SLAM, Graph-based SLAM, and Fast SLAM. These techniques are used in UAS missions for simultaneous mapping and localization of the vehicle in unknown environments. EKF-based SLAM is used for feature-based mapping and has limited scalability. Graph-based SLAM is scalable and can handle large maps. Fast SLAM uses particle filters for efficient and scalable mapping in complex environments.

13. What is the application software used in UAS missions?

Answer. The application software used in UAS missions includes the Robot Operating System (ROS), which provides a flexible framework for building complex robotic systems and has been used extensively to implement autonomous navigation on UGVs and other mobile robotic platforms. Additionally, ROS supports integration with web interfaces through the rosbridge and rosjs client library packages.

14. What is MSDF? Where in a UAS mission is MSDF used?

Answer: MSDF stands for Multi Sensor Data Fusion. It is used in UAS missions during the perception stage to combine and integrate data from multiple sensors to generate a more accurate and comprehensive understanding of the UAS's environment.

15. Give a generic step used in the processing of data obtained from an AUV.

Answer: A generic step used in the processing of data obtained from an AUV is data analysis, which involves the examination and interpretation of the collected data. This step involves filtering, noise reduction, and statistical analyses, which are used to identify trends, patterns, and anomalies in the data. The data analysis step is crucial for generating meaningful insights from the collected data.

16. What are geo-pointing and geolocation functions? How are these achieved by a UAV? What are the applications of geo-pointing and geolocation functions in aerial, ground, and underwater applications?

Answer: Geo-pointing is the process of pointing the camera of a UAV in a specific direction that corresponds to a particular geographic location on the ground. Geolocation is the process of determining the exact geographic coordinates of a target on the ground from the images captured by the UAV's camera. These functions are achieved by using GPS and IMUs that are integrated with the UAV's camera. Applications of geo-pointing and geolocation functions include aerial surveys, search and rescue operations, precision agriculture, and underwater inspections.

REFERENCES

Ashraf, A., Majd, A., et al. (2020). Online path generation and navigation for swarm of UAV's. Hindwai Scientific Programming 8530763.

Caselli, S., Reggiani, M., & Rocchi, R. (2001). Heuristic methods for randomized path planning in potential fields. Proceedings of IEEE Int. Symposium on Computational Intelligence in Robotics and Automation, pp. 426–431. (Stefano, el al, 2001).

Chen, Y. bo, Luo, G. chen, Mei, Y. song, Yu, J. qiao, & Su, X. long. (2014). UAV path planning using artificial potential field method updated by optimal control theory. *International Journal of Systems Science*, *47*(6), 1407–1420. https://doi.org/10.1080/00207721.2014.929191

Engel, J., Sturm, J., & Cremers, D. (2014). Scale-aware navigation of a low-cost quadrocopter with a monocular camera. Robotics and Autonomous Systems, vol. 62, pp. 1646–1656.

Kangunde, V., Jamisola, R.S. Jr., & Theophilus, E.K. (2021). A review on drones controlled in real-time. International Journal of Dynamics and Control, vol. 9, no. 4, pp. 1832–1846. doi: 10.1007/s40435-020-00737-5. Epub 2021 Jan 5. PMID: 33425650; PMCID: PMC7785038.

Kwon, W., Park, J.H., Lee, M., Her, J., Kim, S.-H., & Seo, J.-W. (2020). Robust autonomous navigation of unmanned aerial vehicles (UAVs) for warehouses' inventory application. IEEE Robotics and Automation Letters, vol. 5, no. 1, pp. 243–249. doi: 10.1109/LRA.2019.2955003.

Melo, N., & Lee, J. (2018). Environment aware ADL recognition system based on decision tree and activity frame. Paladyn, Journal of Behavioural Robotics, vol. 9, pp. 155–167. doi: 10.1515/pjbr-2018-0011.

Mohsan, S.A.H., Othman, N.Q.H., Li, Y., et al. (2023). Unmanned aerial vehicles (UAVs): Practical aspects, applications, open challenges, security issues, and future trends. Intelligent Service Robotics, vol. 16, pp. 109–137. https://doi.org/10.1007/s11370-022-00452-4

Mur-Artal, R., & Tardós, J.D. (2017). ORB-SLAM2: An open-source SLAM system for monocular stereo and RGB-D cameras. IEEE Transactions on Robotics, vol. 33, no. 5, pp. 1255–1262.

Newman, P., & Ho, K. (2005). SLAM-loop closing with visually salient features. Proceedings – IEEE International Conference on Robotics and Automation, pp. 635–642. doi: 10.1109/ROBOT.2005.1570189.

Roumeliotis, S.I., Johnson, A.E., & Montgomery, J.F. (2002). Augmenting inertial navigation with image-based motion estimation. *Proceedings 2002 IEEE International Conference on Robotics and Automation (Cat. No.02CH37292)*, Washington, DC, USA, vol. 4, pp. 4326–4333. doi: 10.1109/ROBOT.2002.1014441.

Sabatini, R., Moore, T., & Ramasamy, S. (2017). Global navigation satellite systems performance analysis and augmentation strategies in aviation. Progress in Aerospace Sciences, vol. 95. doi: 10.1016/j.paerosci.2017.10.002.

Samadzadegan, F., & Abdi, G. (2012). Autonomous navigation of Unmanned Aerial Vehicles based on multi-sensor data fusion. 20th Iranian Conference on Electrical Engineering (ICEE2012), pp. 868–873, doi: 10.1109/IranianCEE.2012.6292475.

Sanchez, P. et al. (2020). Real-time collision-free navigation of multiple UAVs based on bounding boxes. MDPI.

Sanchez-Lopez, J.L., Fernandez, S.R., Bavle, H., Sampedro Pérez, C., Molina, M., Pestana, J., & Campoy, P. (2016). AEROSTACK: An architecture and open-source software framework for aerial robotics. *2016 International Conference on Unmanned Aircraft Systems (ICUAS)*, Arlington, VA, USA, 2016, 332–341. 10.1109/ICUAS.2016.7502591.

Wiggers, K. (2020). Waymo's autonomous cars have driven 20 million miles on public roads. Available online: https://venturebeat.com/2020/01/06/waymos-autonomous-cars-have-driven-20-million-miles-on-public-roads/ (accessed on 18 November 2020).

Index

Above ground level (AGL), 118
Accelerometers, 7, 10, 212, 215, 262, 266, 267
Acoustic, 124, 127, 129, 130, 147, 177, 188–192, 205, 208
Aerodynamic, 8, 21, 23, 26, 56, 57, 70, 73, 75, 76, 79, 80, 85, 101, 120, 160
Agriculture, 14, 54, 82, 84–86, 92, 93, 97, 104, 109, 119, 165, 194, 195, 217, 219, 220, 248, 255, 276
Airspace, 13, 21, 84, 89, 90, 93, 94, 106, 108, 115, 116, 149, 153, 156, 259
Algorithm, 34, 55, 61, 65, 82, 98, 123, 127, 134, 144, 146, 154, 160, 162, 167, 181, 189, 213, 215, 216, 222, 225, 226, 233, 239, 246, 247, 251, 254, 255, 257, 259, 261, 263–269, 271, 273, 275
Altitude, 7, 8, 17, 31, 32, 35, 40, 44, 56, 57, 59–61, 75, 79, 83–85, 93–96, 100, 101, 105–107, 110, 113, 114, 118, 121–123, 128, 136, 137, 145, 152, 153, 157, 163, 166, 188, 206, 209, 258, 259, 267
Altitude and Heading Reference System (AHRS), 34
ANSYS, 52
Ant colony optimization, 259
Anti-UAV, 131, 142, 150
Anti-submarine, 174, 187, 192
Archaeological, 92, 93, 97, 164, 187
Archaeology, 158
Artificial intelligence (AI), 146, 156, 213
A-star, 259, 268
Atmosphere, 1
Augmentation system, 60
Auto Navigation System (ANS), 171
Autogyro, 9
Autonomous Navigation System (ANS), 189, 190, 193, 209, 221, 227
Autonomous Underwater Vehicles (AUVs), 16, 168–171, 173, 175, 178, 182, 186, 194, 196, 201, 207, 208
Autonomy level, 108, 116
Autopilot, 30, 60, 61, 91, 194
AUV coordinate system, 196

Backhauling, 41
Bathymetric, 168, 183, 187, 192
Battery eliminator circuit (BEC), 72
Battle Tank, 209
Biometric, 83
Birds of prey, 144
Bluetooth, 35, 38, 43, 46, 47, 63, 133, 213
BMP, 209
Broadband, 39
Buoyancy, 171, 176, 179, 193, 196–200, 202–207
Bureau of Civil Aviation Security (BCAS), 90

Cartography, 158
Catastrophic, 54
Cellular network, 44, 158, 191
Change detection, 162, 163, 164, 182
Characteristics, 42, 44, 67, 69, 70, 73, 75–77, 86, 96, 97, 99–101, 200, 213, 255, 262, 272
Circular error, 54, 55
Civil applications, 93, 118, 169
Clandestine, 177
Classification, 14–18, 89, 97–111, 113–119, 121, 136, 145, 146, 157, 158, 161–164, 168, 169, 178, 192, 223, 271

Classification, 14, 15, 89, 97, 99, 100, 105
Client Library Interface (CLI), 256, 257
Coastline management, 158
Collision drone, 126, 143
Combat Search and Rescue (CSAR), 100
Communication, 2, 3, 7, 8, 10, 14, 15, 17, 18, 22, 23, 26–29, 31, 34–41, 44–47, 56–59, 62–64, 66–68, 89, 92–94, 100, 121, 125, 127–130, 132–134, 146, 147, 152, 153, 159, 164, 168–170, 172–174, 179, 186, 187, 190, 191, 193, 194, 205, 209, 211, 213, 214, 220, 221, 223, 234, 237–239, 246, 247, 250–252, 254, 259–261, 263, 264, 270–273
Communication service, 159
Computational fluid dynamics (CFD), 70
Computer aided design (CAD), 74
Computer vision algorithms (CVA), 269
Constraint diagram, 68, 69, 70, 75, 78
Contour map, 158
Control station (CS), 17, 18, 84, 94, 162, 198, 204, 205, 212, 220, 218
Coriolis, 197, 200, 203, 204, 206, 207
Counter drone, 132, 134, 135, 138, 150, 155
Countermeasures, 141, 148, 183, 187, 192
Counter-UAS (C-UAS), 16, 123, 125, 127, 129, 131, 133, 135, 137, 139, 141, 143, 145, 147, 149, 151, 153, 155
Crowd management, 84, 88
Cruise missile, 99

DARPA, 210, 219
Dazzling, 124, 125, 133, 151–153
Dead reckoning, 171, 190, 258
DEM, 82, 158
Depth, 168, 169, 171, 174, 178–180, 191, 192, 194, 196, 201, 202, 204–206, 207, 212, 215, 257, 258, 265
Design requirement, 86, 88
Diagnostic message, 225, 227
Differential GPS (DGPS), 10
Digital elevation model (DEM), 82
Digital sky, 89
Digital surface model (DSM), 82, 244
Digital terrain model (DTM), 82, 158
Dijkstra, 259, 268
Direct reckoning, 11
Director General of Civil Aviation (DGCA), 88
Disaster management, 40, 82, 92, 93, 104, 119, 165
Do-it-Yourself (DIY), 7
Doppler Velocity Lag (DLG), 34, 180, 191
Downlink, 3, 22, 23, 29, 30, 44, 59, 60
Drive-by-Wire (DBW), 209, 212, 213, 226, 227, 231–234, 236, 242, 250
Driving mode, 226, 227
Drone, 1, 3–8, 14, 16, 17, 19, 20, 25, 43, 46, 49, 51–54, 57, 64–66, 82, 85–93, 97, 98, 101, 109, 111–113, 116, 117, 120, 122–139, 141, 143, 144, 146–162, 167, 177, 181, 185, 195, 265, 175
Drone bullet, 143
Drone detection, 122, 123, 130, 139, 148, 155
Drone dome, 133–136, 150
DTM, 82, 158
Dull, Dirty, or Dangerous (DDD), 12
Dyadic communication, 172

Earth Centered Earth Fixed (ECEF), 171
Economic reasons, 12, 13
Electro magnetic spectrum (EMS), 15
Electro optical (EO), 124, 134
Electromagnetic compatibility (EMC), 28, 255
Electromagnetic interference (EMI), 28, 255
Electromagnetic signature (EMS), 53
Electromechanical actuation (EMA), 231
Electronic attack (EA), 100, 134, 135
Electronic counter measures (ECM), 163
Electronic speed control (ESC), 71
Embedded computer, 233, 238, 250, 251
Embedded software, 232, 238, 239, 246, 251
Empennage, 67, 72–74, 76, 77, 80
Endurance, 7–9, 15, 17, 18, 21, 23, 27, 67, 72, 75, 85–88, 101, 104, 107, 110, 112, 114, 146, 156, 157, 163, 169, 170, 177, 178, 193, 231, 241, 242, 252, 255, 273
Enforcement, 4, 18, 82, 83, 90
Environmental, 12, 13, 18, 19, 82, 86, 88, 92, 94, 96, 97, 112, 117, 119, 141, 146, 149, 163, 165, 168, 177, 181, 186, 189, 191, 192, 200, 204, 207, 213, 226, 234, 249, 251–253, 255, 258, 259, 266, 273
Explosive ordnance disposal, 211, 248
External device (ED), 45

Fail safe, 88, 166, 233, 240, 259, 270, 272
False negative, 128, 148, 152
False positive, 128, 148, 152
Federal Aviation Administration (FAA), 99
Field of View (FoV), 29, 54, 56, 83, 98, 135, 136, 144, 148, 159, 161, 164, 258
Finite element method (FEM), 52, 73
Flood modelling, 158
Flying Ad Hoc Network (FANET), 38, 41, 46
Fog bomb, 128, 152
Frequency division duplexing (FDD), 39
Functional requirement, 85, 88, 251
Fuselage, 57, 67, 72–74, 76–78, 80

Galileo, 33, 130, 188
Generic classification, 101, 102
Geolocation, 29, 59, 60, 275
Geological, 168
Geo-pointing, 29, 59, 60, 275
Gimbal, 23, 25, 56, 86, 88, 136, 167
Global navigation satellite system (GLONASS), 30, 33, 125, 130, 171, 188, 237, 250, 258
Global positioning system (GPS), 5, 10, 30, 31, 33, 34, 54, 57, 62, 63, 89, 93, 95, 124, 129, 135, 142, 152, 159, 163, 166, 171, 188, 213, 219, 258, 262
GNSS jamming, 124, 126, 129, 133, 150, 267
Goable, 258
Graphical user interface (GUI), 26, 211, 220, 227, 229, 240, 241, 252, 257
Ground control system (GCS), 18, 23, 58
Ground pilot, 3, 17, 18, 23, 26, 27, 53, 59, 65, 156, 270
Gyro-stabilized, 22, 136

HELIOS, 133, 134, 150
Heterogeneous networks (HetNets), 8
High-altitude long-endurance (HALE), 7, 15, 102, 107, 114, 163, 164
High Endurance Autonomous Underwater Vehicle (HEAUV), 178
Holographic Radar, 144, 145
Hull design, 201, 205
Human-Robot Interface (HRI), 214, 215
Hydrodynamic damping, 197
Hydrofoils, 179
Hydrographic, 168, 174, 192
Hyperspectral, 96, 165, 175, 195

Identification, 75, 85, 89, 106, 108, 115, 122, 124, 127, 134–137, 152, 155, 161, 181, 208, 210, 239, 240, 271
Imaging and mapping, 159
Impact test, 51
Inertial Navigation Systems (INS), 10, 30, 32, 33, 34, 61, 159, 163, 171, 180, 181, 188, 190, 237, 250, 257, 258, 262, 266, 267
Infrared (IR), 56, 123
Instantaneous field of view (IFoV), 159, 161, 164
Intelligence, reconnaissance and surveillance (ISR), 18, 134, 135, 162, 174
Interdiction, 124–126, 128, 129, 133, 146, 149, 151–153
Interdiction effectiveness, 129, 149
Interdiction hazard, 128, 149, 153
Interface hardware, 234
International Civil Aviation Organization (ICAO), 4
IR emitting beacon, 56
IRNSS, 212, 237, 250, 258
ISTAR, 14, 16, 156, 157, 159–161, 163–165, 167

Jamming, 31, 34, 124–126, 128–130, 133, 136, 141, 146, 148, 150, 152, 153, 266, 267
Joystick, 26, 213, 214, 246, 247, 251

Keyboard Video Mouse (KVM), 238, 240, 252
Knowledge graph, 21, 24

Landing gear, 5, 23, 47–53, 55, 56, 64, 67, 77, 78
Landing type, 99, 100, 103
Lateral load transfer (LLT), 70
Legged locomotion, 212
Legitimate, 128, 148–151, 153
LiDAR mapping, 81, 82, 85, 92
Light Detection and Ranging (LiDAR), 81, 189, 257
Li-ion, 71
Line of actual control (LAC), 90
Line of control (LOC), 90
Line of sight (LoS), 14, 55–56, 107, 108, 114, 115, 159, 172, 213, 260, 273
Location services, 159

LOCUST, 133–135, 150
Long term evolution (LTE), 39, 63
Low Earth Orbit (LEO), 44

Magnetic compass, 34, 191
Medium-altitude long-endurance (MALE), 7, 15, 102, 107, 110, 114, 163, 164
Mapping, 14, 15, 81, 82, 85, 92, 96, 97, 112, 113, 119, 158–161, 163, 165, 174, 175, 178–179, 181, 182, 186, 187, 192, 204, 212, 215, 224, 226, 244, 246, 247, 249, 257, 274
Mars exploration rover, 218
Maximum gross take-off weight (MGTOW), 105, 106, 118
Maximum take-off weight (MTOW), 17, 69, 99–102, 106, 109, 110
Mechatronic I, 223, 224
Microcontrollers, 7, 261
Microprocessor, 26
Microwave, 124, 125, 128, 136, 137, 152, 154, 155, 244
Microwave proximity sensor (MPS), 244
Middleware, 255, 256, 272, 273
Military, 7, 9, 12, 14, 18, 44, 99, 120, 147, 160, 168, 169, 177, 178–179, 186, 211, 224, 255
Military standard (MILSTD), 133
MIMO, 66, 130
Minefields, 168, 212
Mineral, 97, 158, 168, 218
Mission configuration, 226, 228
Mission control software (MCS), 257, 272
Mission control system (MCS), 227, 238, 252, 257, 272
Mission objectives, 1, 9, 21, 117, 169, 174, 188, 192, 205, 254, 259, 262, 271
Mission planning and control system (MPCS), 22
Mission-specific payload, 211, 214, 220
Modularity, 201, 230, 255, 272
Moving geometry, 21, 67, 196
Multispectral, 17, 86, 93, 96, 98, 175

Navigation and guidance, 34
Navigation system, 1, 2, 5, 9, 10, 17–19, 30, 31, 33, 57, 67, 142, 150, 151, 153, 159, 171, 188–190, 193, 206, 209, 215, 221, 224, 227, 230, 258, 263, 266, 267
Ni-Cd, 71
Ni-MH, 72
NMEA, 34, 61, 62, 258
No-fly zone, 88, 90
No permission no take-off (NPNT), 106, 108, 115
Non-line-of-sight (NLoS), 158, 159, 161, 164, 173, 213, 233, 238, 260
Nuclear biological and chemical (NBC), 244

Observe Orient Decide and Act (OODA), 160–162, 164
Obstacle limitation surfaces (OLS), 90
Oil and gas exploration, 158
Operating system (OS), 46, 250, 254, 256, 271
Operator permit, 108
Opportunistic communication, 172
Optical, 22, 23, 55, 124, 130, 131, 134, 136, 147, 154, 174, 182, 190, 192, 258
Orthophoto, 158
Overwatch, 141, 142, 150

Parachute, 52, 53, 57, 128, 138–141
Photogrammetry, 19, 81, 82, 85, 92, 97, 98, 156–161, 163–165, 167, 182, 192, 195
Pitbull, 141, 142
PHASER, 136, 137, 150
Platform-mounted display (PMD), 238, 240, 242, 252
Precipitation, 23
Precision landing, 53–56, 65
Point-to-point, 39
Potentiometer, 26
Propulsion design, 201
Projectile, 124–126, 162
Pulse per second (PPS), 34
Planimetric, 158
Pollution modelling, 158
Polynomial Affine Transform (PAT), 161
Photogrammetric functions, 92, 160, 163, 164
Para-dropping, 170
Proximity, 142, 174, 175, 244, 258
Pollution, 13, 23, 104, 119, 158, 163, 193
Payload, 1, 2, 9, 17, 18, 20–23, 26–29, 34, 35, 56–60, 63, 67, 68, 72–74, 77, 81, 82, 85–88, 93, 95, 96, 101, 104, 113, 137, 138, 140, 146, 149, 157, 160, 166,

168, 169, 173, 175, 179, 191–193, 205, 211, 214, 220, 222, 223, 226, 231, 245, 248, 254, 255, 260, 270–273
Propulsion, 17, 21, 23, 56, 57, 67, 70, 71, 74–78, 80, 95, 96, 99–101, 103, 109, 111, 112, 120, 140, 169, 170, 179, 186, 195, 201–203, 205, 222
Propeller, 5, 9, 15, 21, 23, 56, 57, 68, 70–72, 78, 87, 88, 91, 95, 12, 139, 141, 143, 170, 179, 195, 201, 205
Propulsion system, 23, 57, 67, 70, 103, 111, 202, 203, 205

Radar, 10–12, 17, 22, 23, 25, 56, 97, 109, 121, 123, 127, 129–131, 134–137, 139, 144, 145, 147, 150, 152, 153, 155, 160, 174, 194, 212, 215, 220–222, 236, 248, 249, 258, 267, 269
Radio, 7, 10–11, 30, 35–37, 57, 62, 91, 95, 123, 127, 130, 132, 136, 137, 147, 152, 171, 180, 188, 189, 191, 210, 213, 219, 247, 263, 267
Radio frequency (RF), 123, 124, 136, 138, 147, 152, 171, 188, 189, 263
Real time kinematics (RTK), 33
Rechargeable batteries, 112, 179, 253
Reconnaissance, 12, 18, 99, 100, 120, 134, 156, 187, 248
Reduced instruction set computing (RISC), 45
Remote control, 25–27, 58, 59, 130, 211, 247
Remote desktop protocol (RDP), 45
Remote operation, 220, 221, 233
Remotely operated underwater vehicle (ROUV), 177
Remotely piloted aircraft (RPA), 3–6
Remotely piloted vehicle (RPV), 116
Retractable landing gear, 47, 50
RF direction finding, 135
RF jamming, 124, 126, 129
RF signature, 130
Righting momentum (RM), 199
Rigid Hull Inflatable Boats (RHIB), 177–178
Robotics Operating System (ROS), 256, 257
Robustness, 256, 274
Ruggedized, 28

Salinity, 168, 175, 179, 186, 192, 193, 206
Satellite communication (SATCOM), 3, 37, 41, 44, 63
Scalability, 255, 268, 272, 274
Scale Invariant Feature Transform (SIFT), 55, 66
Seabed, 168, 181
Secure environment, 26
Security, 9, 19, 20, 66, 82–84, 87, 90, 98, 108, 115, 116, 119, 120, 126, 131, 139, 147, 150, 152, 154, 155, 167, 187, 251, 255, 259, 272, 275
Seismic data, 168
Sensor payload, 1, 2, 17, 21, 168, 169, 175, 191
Sentinel catch, 138–140, 150
Sequence of waypoints, 229
Servo, 26, 68, 72, 91, 119
Simultaneous Localization and Mapping (SLAM), 98, 160–164, 181, 194, 215, 244, 245, 257, 258, 264–267, 271, 274–276
Size Weight Area and Power (SWAP), 137, 138, 156, 160, 163, 164, 243, 254, 256, 260, 273
Slingshots, 126, 152
Small Unmanned Aerial Systems (SUAS), 100
Society of Automotive Engineers (SAE), 108
Solar power, 103, 112, 179
SPARROWHAWK, 137, 138, 150
Spherical error, 54
Spoofing, 30, 31, 34, 59, 60, 124, 125, 129, 132, 148, 150–153, 163, 164
Stability, 21, 67, 70, 73, 75–78, 80, 95, 113, 159, 166, 196–200, 204, 207, 208, 212
Stability of the AUV, 196
Strategic, 83, 89, 90, 107, 114, 154, 209
Submarine, 171, 177, 187, 192, 193, 201
Submerging, 201, 202, 205, 208
Surveillance, 8, 12–14, 18, 19, 35, 40, 44, 54, 58, 82, 84, 87, 92, 93, 97, 100, 104, 112, 119, 120, 131, 134, 136, 147, 149, 153, 156, 158–161, 163, 164, 167, 168, 174, 187, 192, 193, 211, 214, 217, 223, 226, 247, 263
Swarm, 8, 19, 36, 66, 105, 117, 131, 133, 136, 141, 154, 169, 254, 275
Synthetic Aperture Radar (SAR), 174, 194

Teleoperation, 210, 214, 227–230, 235, 236, 240, 247, 249, 253
Temporary reserved airspace (TRA), 90

Thermal image, 23
Time division duplexing (TDD), 39
Time To First Fix (TTFF), 34
Topography, 158, 174
Torpedoes, 168
Tracking, 10, 11, 18, 20, 44, 55, 83, 87, 88, 119, 121–124, 127, 136, 141, 146, 147, 152, 155, 159, 162, 167, 187, 192, 194
Traffic management, 84, 88, 93, 239
Transmission systems, 2, 17
Transport modelling, 94, 248
Triangulation, 142, 162, 171
Turbidity, 168
Twin boom, 73

Ubiquitous, 40, 41
UCAV, 8, 163, 164
Uncrewed Marine Vehicle (UMV), 177
Under Water Surveillance, 14–15
Unexploded bombs (UXB), 9
Unique identification number (UIN), 89
Unmanned aerial operators permit (UAOP), 106, 108
Unmanned aerial vehicle (UAV), 1–6, 7, 9, 11–13, 15–19, 21–36, 39–41, 44
Unmanned aircraft systems (UASs), 22, 154, 276
Unmanned autonomous systems (UAS), 1–6, 8–10, 12–17, 19, 22, 34–37, 99, 110, 106–108, 114, 115, 118, 121, 122, 126–128, 134, 136, 144, 145, 149, 156, 167, 254–260, 262–270
Unmanned combat air vehicles (UCAVs), 8, 163, 164
Unmanned ground vehicle (UGV), 15, 16, 32, 34, 60, 171, 190, 209–235, 237, 239–250, 252–254, 256, 258, 260, 270, 271, 274
Unmanned surface vehicle (USV), 32, 60
Unmanned underwater vehicle (UUV), 194
Urban planning, 82, 158, 163

Virtual energy method, 51
Vision based navigation, 180, 188, 189
Volumetric survey, 158
VTOL, 91, 95, 103, 112, 113

Wartime drone, 122
Waypoint, 30, 57, 60, 66, 86, 116, 216, 225, 227, 229, 230, 241
Weather, 10, 12, 22, 56, 93, 123, 127, 130, 146, 189, 213, 215, 243
Wi-Fi, 37
Wing design, 69, 77
Wireless, 30, 35, 37–39, 43, 44, 45, 59, 60, 63, 66, 89, 185, 190, 191, 210, 213, 219, 233, 234, 238, 241, 247, 252, 253, 260, 273
WLAN, 36
World coordinate system, 171

Zigbee, 37–39, 43, 63, 213